US MILITARY ENCYCLOPEDIAS

THE MILITARY VEHICLES ENCYCLOPEDIA

BY ARNOLD RINGSTAD

Encyclopedias

An Imprint of Abdo Reference
abdobooks.com

TABLE OF CONTENTS

US MILITARY VEHICLES 4
A-10 Thunderbolt II 6
AAV-7A1 Assault
   Amphibious Vehicle 8
AC-130J Ghostrider 10
AH-1Z Viper 12
AH-6 Little Bird 14
AH-64 Apache 16
America-Class Amphibious
   Assault Ship 18
Arleigh Burke–Class Destroyer 20
AV-8B Harrier II 22
Avenger-Class Mine
   Countermeasures Ship 24
B-1B Lancer 26
B-2 Spirit 28
B-52H Stratofortress 30
Buffalo Mine-Resistant Ambush
   Protected (MRAP) Vehicle 32
Bv 206 34
C-5M Super Galaxy 36
C-17 Globemaster III 38
C-130J Hercules 40
CH-47F Chinook 42
Cougar MRAP 44
E-2 Hawkeye 46
E-3 Sentry 48
E-8 Joint STARS 50
EA-18G Growler 52
F-15C Eagle 54
F-15E Strike Eagle 56
F-16C Fighting Falcon 58
F-22A Raptor 60
F-35 Lightning II 62
F/A-18 Hornet 64
F/A-18 Super Hornet 66

Family of Medium
   Tactical Vehicles (FMTV) 68
Famous-Class Cutter 70
Freedom-Class Littoral
   Combat Ship (LCS) 72
Gerald R. Ford–Class
   Aircraft Carrier 74
Global Positioning System (GPS)
   Satellites 76
HC-144 Ocean Sentry 78
Healy-Class Icebreaker 80
Heavy Expanded Mobility
   Tactical Truck (HEMTT) 82
Humvee 84
Husky Mounted Detection System
   (HMDS) 86
Independence-Class LCS 88
Joint Light Tactical Vehicle (JLTV) 90
Juniper-Class Buoy Tender 92
KC-10 Extender 94
KC-46A Pegasus 96
KC-135 Stratotanker 98
Keeper-Class Buoy Tender 100
Light Armored Vehicle (LAV-25) 102
Legend-Class Cutter 104
Lewis B. Puller–Class
   Expeditionary Sea Base (ESB) 106
Logistics Vehicle System
   Replacement (LVSR) 108
Los Angeles–Class
   Attack Submarine 110
M1 Abrams 112
M2 Bradley 114
M9 Armored Combat
   Earthmover (ACE) 116
M88A2 HERCULES 118

M104 Wolverine Armored Bridgelayer 120
M113A3 Armored Personnel Carrier (APC) 122
M160 Robotic Mine Flail 124
M1070 Heavy Equipment Transporter 126
M1126 Stryker Combat Vehicle 128
M1150 Assault Breacher Vehicle 130
Marine Protector–Class Coastal Patrol Boat 132
Medium Tactical Vehicle Replacement (MTVR) 134
MH-65E Dolphin 136
MQ-1C Gray Eagle 138
MQ-9 Reaper 140
MRZR Alpha Light Tactical Vehicle (LTV) 142
MV-22B Osprey 144
Nimitz-Class Aircraft Carrier 146
Ohio-Class Submarine 148
P-8A Poseidon 150
R-11 Aircraft Refueler 152
Reliance-Class Cutter 154
Response Boat-Medium (RB-M) 156
Response Boat-Small (RB-S) II 158
RQ-4 Global Hawk 160
RQ-11B Raven 162
RQ-20B Puma 164
Sentinel-Class Cutter 166
Space-Based Infrared System (SBIRS) Satellites 168
T-6A Texan II 170
T-38C Talon 172
Ticonderoga-Class Cruiser 174
UH-1Y Venom 176
UH-60M Black Hawk 178
Virginia-Class Attack Submarine 180
Wideband Global SATCOM (WGS) Satellites 182
X-37B Orbital Test Vehicle 184
Zumwalt-Class Destroyer 186

GLOSSARY 188
TO LEARN MORE 189
INDEX 190
PHOTO CREDITS 191

US MILITARY VEHICLES

The Humvee is one of the most common land vehicles used by the US military.

The US military is one of the most capable fighting forces on the planet. A key part of its strength comes from the vast assortment of vehicles it uses. High-tech machines that drive on land, fly through the air, and cruise on the sea give the military its unrivaled capabilities.

TYPES OF VEHICLES

On land, the military's armored vehicles, such as tanks and infantry fighting vehicles, provide soldiers with protection and support. Instead of wheels, these vehicles drive on tracks, rumbling over rough terrain to reach the front lines. Powerful trucks haul weapons and cargo into battle. Support vehicles carry damaged tanks out of the fight to be repaired. Engineering vehicles clear minefields, build bridges, and more.

In the air, the military's fighter jets clear the skies of opposing aircraft. Bombers strike with pinpoint accuracy at enemy bases and infrastructure. Transport planes deliver soldiers, supplies, and more to battlefields around the globe. Drones give military leaders a clear view of the fighting below.

Small aircraft, such as the T-6A Texan II, are able to fly in tight formations.

Helicopters provide transportation and close air support to troops on the ground. High above Earth, satellites provide vital intelligence and communications.

At sea, the military's aircraft carriers bring powerful jets thousands of miles away from friendly shores. They are protected by destroyers and cruisers. Amphibious vessels carry firepower from the sea to the shore. Beneath the waves, hard-to-detect submarines may be used to hunt enemy ships. Along US coastlines and waterways, the US Coast Guard's cutters defend the country, enforce the law, and maintain safe travel throughout US waters.

The vehicles of the US military are important tools. But many of their capabilities come from being operated by highly trained soldiers, pilots, sailors, and other military professionals. This combination of technology and skill is what gives the US military its unrivaled power.

A Nimitz-class aircraft carrier has approximately 6,000 crew members who manage the warship and aircraft on board.

A-10 THUNDERBOLT II

Many Air Force jets support troops by attacking ground targets. But the A-10 Thunderbolt II is the first aircraft designed specifically for this job. Pilots often call this tough plane the Warthog.

This airplane is a single-seat jet. Titanium armor surrounds the cockpit to protect the pilot. Backup systems help it keep flying even if the plane takes damage. It is able to take off and land on rough surfaces, such as dry lake beds. This allows the Thunderbolt II to fly near the front lines.

The Thunderbolt II was designed around its fearsome main weapon. The GAU-8/A Gatling gun can fire 4,200 rounds a minute. Each of these 30-mm rounds weighs more than 1.5 pounds (0.7 kg). They are designed to destroy armored vehicles, including tanks. The Thunderbolt II can also carry 16,000 pounds (7,260 kg) of other weapons, such as missiles and bombs.

HISTORY

The Thunderbolt II entered service in 1976. It saw heavy use during the Persian Gulf War (1990–1991). This type of jet flew more than 8,000 missions, destroying hundreds of tanks and other vehicles.

The Thunderbolt II reaches a top flying speed of about 420 mph (676 kmh).

Thunderbolt II jets line up in a formation known as an Elephant Walk in preparation for takeoff.

AAV-7A1 ASSAULT AMPHIBIOUS VEHICLE

Marines use the AAV-7A1 Assault Amphibious Vehicle to land ground troops on enemy coastlines. The vehicle operates on both water and land. Water jets propel the vehicle forward at sea. On land, the vehicle moves on tracks like a tank. It drives ashore and supports troops as they move inland. This armored vehicle is commonly known as an amtrac, short for "amphibious tractor." It can reach speeds of up to 45 miles per hour (72 kmh) on land and 8.2 miles per hour (13.2 kmh) in the water.

Amphibious vehicles are designed to move both on land and at sea.

Amphibious vehicles played a major role in World War II (1939–1945) and are still important US Marine Corps vehicles today.

TYPES OF AAV-7A1

There are three different versions of the AAV-7A1. The most common is called the Personnel version. It carries 25 Marines, along with a crew of four to control the vehicle. An M2 machine gun and Mark-19, or Mk-19, automatic grenade launcher are used to defend the vehicle and support ground troops. The Command version carries commanding officers, radio operators, and extra radio equipment to communicate with other units. The Recovery version comes with a crane, tools, and equipment for repairing damaged vehicles.

AC-130J GHOSTRIDER

A Ghostrider has four large engines that spin its propellers.

The AC-130J Ghostrider entered service with the Air Force in the 2010s. It is based on the C-130 transport plane. These older transport planes were used heavily during the Vietnam War (1954–1975).

WEAPONRY

Gunships are large aircraft with powerful weapons that fire from the sides. They can circle ground targets, attacking from high above. The Ghostrider's high-tech navigation and sensor systems let it strike precise locations. This helps reduce collateral damage in urban areas.

The Ghostrider is equipped with a 30-mm cannon and a 105-mm cannon. The 30-mm cannon fires up to 200 rounds per minute. Each round is about the size of a soda bottle. The 105-mm cannon shoots rounds that are about 3 feet (0.9 m) long and weigh 50 pounds (23 kg). The gunship can also launch guided missiles, such as the Hellfire and Griffin. The crew uses video screens and control sticks to aim and fire these weapons.

Members of the Air Force operate the 105-mm cannon of a Ghostrider gunship as part of a training exercise.

AH-1Z VIPER

The AH-1Z Viper is an attack helicopter flown by the US Marine Corps. It entered service in 2011. The rugged helicopter has two engines and can still fly if one engine is damaged.

The Viper has a crew of two. The pilot sits in the front, and the gunner sits in the back. Each wears a helmet-mounted display. This keeps important information, such as speed and

The blades of the AH-1Z Viper span 48 feet (14.6 m) from tip to tip.

AH-1Z Vipers are designed to provide air support for troops on the ground.

altitude, in view at all times. The helmet can also switch to night-vision mode for missions at night.

WEAPONRY

The Viper can carry a wide variety of weapons. TOW missiles and Hellfire missiles destroy tanks. Hydra guided rockets strike other ground targets. Sidewinder missiles shoot down aircraft. Finally, a 20-mm Gatling gun provides even more firepower. The gunner can aim it using his or her helmet. Simply looking at the target with the helmet on will point the gun in the correct direction.

AH-6 LITTLE BIRD

Soldiers train so they can quickly enter and exit a Little Bird while it is in flight.

The Little Bird gets its name from the fact that it is relatively small compared with other helicopters. Despite its size, it is a deadly and effective vehicle. It comes in two versions. The AH-6 is an attack helicopter loaded with weapons. The MH-6 is used for transporting troops.

The Little Bird was originally designed for Army scouting missions. However, it has mainly been used by special forces. The Army's 160th Special Operations Aviation Regiment (SOAR) uses the helicopter for its missions. The helicopter's small size allows it to deliver special forces troops onto rooftops or into tight spaces in urban areas.

CREW AND WEAPONRY

The helicopter has a crew of two and can carry up to six passengers. Some of them sit outside on seats attached to the side doors. The AH-6 carries a minigun, rocket launchers, and guided missiles, providing powerful support for troops on the ground.

An AH-6 Little Bird fires a rocket during a training exercise.

AH-64 APACHE

The AH-64 Apache is an attack helicopter used by the Army. It has a crew of two, consisting of the pilot and a gunner. Its mission is to destroy ground targets, including tanks, lighter vehicles, and soldiers. The helicopter's powerful weapons include Hellfire missiles, Hydra rockets, and a 30-mm cannon. It has strong defenses too. An armored cockpit protects the crew from machine gun fire. The exhaust system reduces engine heat. This makes it harder for heat-seeking missiles to detect the aircraft.

The AH-64 Apache is the US military's primary helicopter for offensive combat.

Regular maintenance and inspections ensure that aircraft are fit for missions.

ADVANCED TECHNOLOGY

The AH-64E version of the Apache was introduced in 2011. It includes advanced communications, navigation, and sensor systems. Its radar can track multiple targets at once, and a laser targeting system helps the crew fire weapons accurately. The crew can even control a drone from the cockpit. The video feed from the drone is visible on the crew's screens. This improves the crew's awareness of events on the battlefield.

AMERICA-CLASS AMPHIBIOUS ASSAULT SHIP

An MV-22B Osprey prepares to land on the deck of the USS *America*.

The Navy's America-class vessels are amphibious assault ships. These ships carry jets and helicopters. For many missions, the Navy works alongside the Marines. The mission of the America-class ships is to quickly and safely move Marines and their equipment onto an enemy coastline. Then, the aircraft from the amphibious assault ship provide support for the troops on the shore. The first America-class ship, the USS *America*, entered service in 2014.

CREW AND WEAPONS

America-class ships have a crew of about 1,000 sailors. They can also carry about 1,700 troops. The ships are equipped with several kinds of weapons to defend themselves. Missile launchers and machine guns fire at enemy sea and air targets. The Phalanx close-in weapon system (CIWS) detects incoming threats, such as missiles or aircraft, and shoots them down with a rapid-firing machine gun.

The USS *America* can carry about 20 small aircraft at once.

ARLEIGH BURKE–CLASS DESTROYER

Destroyers are naval ships that are designed for combat. They have weapons that can take out aircraft and submarines. Some Arleigh Burke–class destroyers can also launch missiles at targets on land.

This class of ships was named for a Navy officer who commanded destroyers in the Pacific during World War II (1939–1945). The Arleigh Burke class entered service in 1991. Since then, the ships have been upgraded many times. These destroyers can work on their own, or they can join with groups of other Navy ships.

CREW AND WEAPONS

Each ship has a crew of about 350 people. The destroyer's weapons include Mk-46 torpedoes, a 5-inch Mk-45 artillery gun, and Sea Sparrow missiles to shoot down

The USS *Arleigh Burke* is about 505 feet (154 m) long.

Sailors stand on the deck of the USS *Arleigh Burke* as they wait for a cargo ship to arrive with supplies.

enemy aircraft. It also features the Mk-41 Vertical Launching System (VLS). The VLS holds dozens of missiles. This includes Tomahawk cruise missiles, which are used against ground targets. Newer versions of the ships in this class also carry two helicopters. These helicopters search for submarines and attack them using torpedoes.

AV-8B HARRIER II

The AV-8B Harrier II is a ground-attack jet used by the Marine Corps. It is capable of vertical or short takeoffs and landings (V/STOL). This means the jet can point its thrust downward to lift straight off the ground. V/STOL gives the Harrier II a lot of flexibility. Pilots can fly it from aircraft carriers, amphibious assault ships, and small bases on land.

HISTORY

The Harrier II has a long history. In 1957, the British designed the Kestrel fighter jet. It was one of the first jets capable of V/STOL. The same company developed this into the original Harrier in the 1960s. The Harrier II followed in 1978. Cooperation between the United States and the United Kingdom brought the Harrier II to the US Marine Corps.

This jet has a single pilot. Its weapons include two 25-mm cannons. It also carries thousands of pounds of bombs and missiles. In more than four decades of service, it has proven to be a flexible and powerful jet.

The AV-8B was one of the first US military planes capable of V/STOL.

Two Marines refuel a Harrier II.

AVENGER-CLASS MINE COUNTERMEASURES SHIP

Militaries place naval mines to block important waterways. These explosive devices prevent ships from carrying out their missions. To remove these mines, the US Navy uses Avenger-class mine countermeasures ships. These vessels entered service in 1987. Each ship has a crew of about 80 people.

TECHNOLOGY AND WEAPONS

Avenger-class ships have several systems that let them find and destroy mines.

In 2020, the US military had 11 Avenger-class mine countermeasures ships in service, including the USS *Dextrous*.

Crew members of the USS *Scout*, an Avenger-class ship, detonate an explosive device from a safe distance during a mine countermeasures training exercise.

Sonar equipment sends out pulses of sound to detect mines. Underwater video cameras let the crew see exactly where mines are. Cable-cutting devices free mines that are anchored to the seafloor. And detonating devices allow crew members to safely explode the mines they find.

The ships have hulls made of wood and a special plastic. This makes them less likely to set off magnetic mines. These tough materials can also withstand an explosion from a nearby mine. Avenger-class ships are not designed for battle, but they do carry weapons to defend themselves. Each ship has several machine guns and two grenade launchers.

B-1B LANCER

The Air Force's B-1B Lancer is a long-range bomber aircraft. One of its notable features is its moving wings. They stick straight out to the sides at slow speeds, such as during takeoff and landing. They sweep back for high-speed flight.

The concept for the B-1 was developed in the 1970s. The design was for a fast, high-altitude bomber. But the program was canceled in 1977. The US military revived it in 1981 as the B-1B. The new version was not designed to fly as quickly, but it had better radar and could carry more weight. The B-1B entered service in 1986.

The B-1B Lancer is designed to reach speeds of more than 900 miles per hour (1,450 kmh).

The B-1B can fly 4,600 miles (7,400 km) before it needs to be refueled.

CREW AND WEAPONS

The B1-B bomber has a crew of four. The commander and copilot fly the plane. Two combat systems officers handle the weapons. The Lancer was originally designed to carry nuclear weapons. But treaties limited the number of nuclear bombers. Today the Lancer carries a mix of nonnuclear bombs, mines, and missiles. For example, a single B-1B can carry 84 Mk-82 bombs that weigh 500 pounds (227 kg) each.

B-2 SPIRIT

The B-2 Spirit is a bomber flown by the Air Force. This plane looks different from most other military aircraft. It has a flying-wing design, meaning its body and wings blend together into a single shape. It has no separate tail. This design reduces drag, allowing the plane to fly more efficiently.

The flying-wing design also helps with the bomber's most important feature: stealth. In addition to its shape, the plane's materials and coatings help it hide from enemy radar. This allows it to sneak through enemy air defenses and destroy challenging targets. Many details of the stealth technology are top secret.

A B-2 Spirit flies behind a tanker aircraft to be refueled in flight.

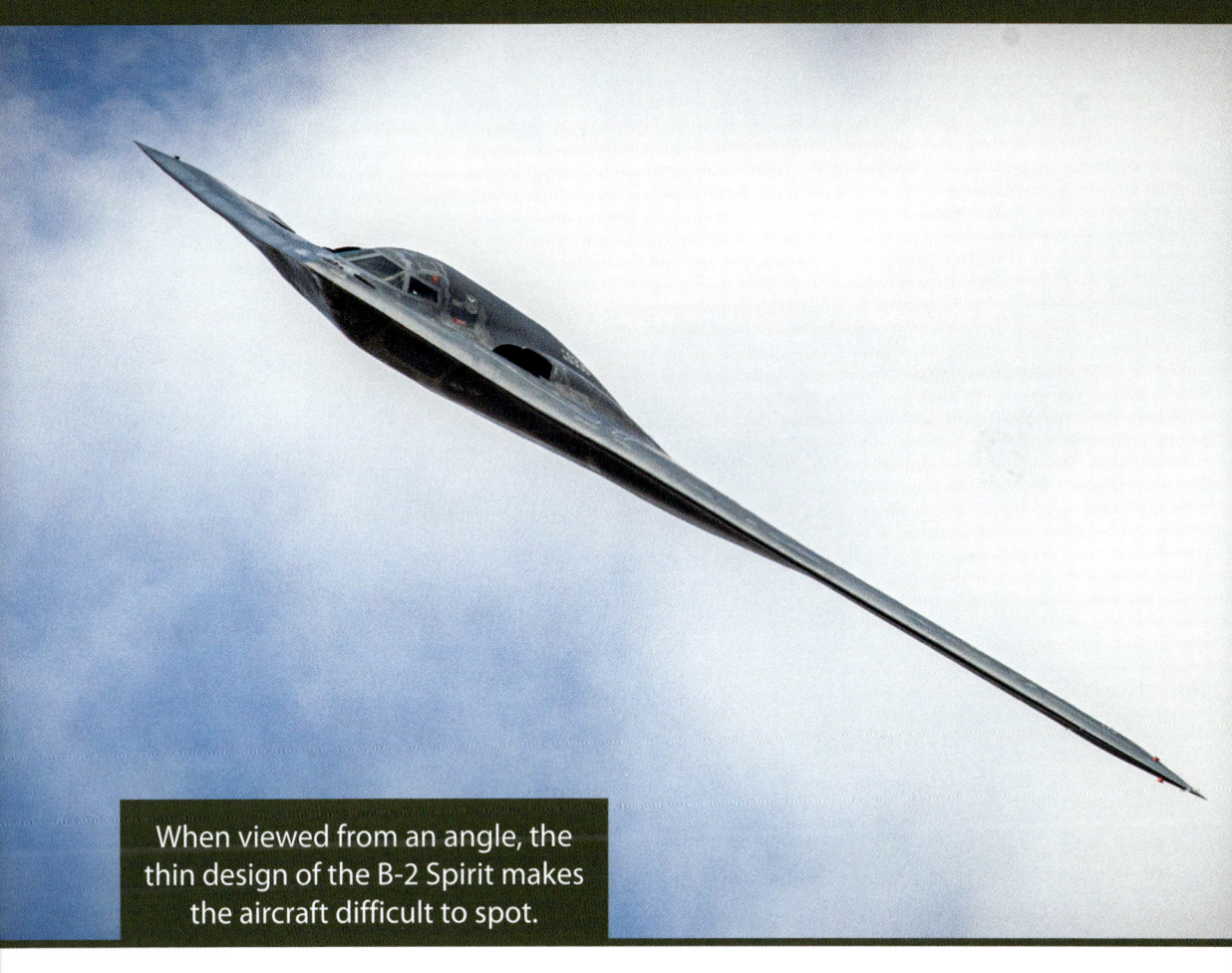

When viewed from an angle, the thin design of the B-2 Spirit makes the aircraft difficult to spot.

HISTORY AND USE

The B-2 was revealed to the public in 1988, and it entered service in 1993. It has flown many missions since then. The bomber can fly for about 6,000 nautical miles (11,110 km) before it needs to be refueled. This can be done on land or in the air. In the Afghanistan War (2001–2014), aerial refueling allowed some B-2s to fly nonstop missions from Missouri to Afghanistan.

B-52H STRATOFORTRESS

The B-52H Stratofortress can carry about 70,000 pounds (31,750 kg) of weapons.

The B-52H Stratofortress is a long-range heavy bomber. The Air Force uses it to carry both nuclear and nonnuclear bombs. It can also launch cruise missiles. The jet can be used to help the Navy attack ships too.

The B-52H is a large aircraft, with a wingspan of 185 feet (56 m). It is powered by eight jet engines arranged in four pairs on its wings. It has a crew of five. In addition to the commander

and pilot, there are two navigators and an operator for electronic equipment.

HISTORY TO TODAY

This bomber has a long history with the Air Force. The earliest version first flew in 1954. The current model, the B-52H, entered service in 1961. Modern B-52H bombers have much more advanced technology than the original planes. Targeting pods allow the crew to identify, track, and strike ground targets. B-52H jets have been used in recent conflicts, such as the fight against the terrorist group ISIS in Syria.

B-52H Stratofortresses are closely inspected by maintenance workers each month to make sure the aircraft are safe to fly.

BUFFALO MINE-RESISTANT AMBUSH PROTECTED (MRAP) VEHICLE

The Buffalo is a mine-resistant ambush protected (MRAP) vehicle. It was originally developed in South Africa. It was then sold to the United States and other countries. The US Army uses the Buffalo to clear paths through areas that may have mines. The vehicle has a large robotic arm. The crew can control the arm to investigate and move mines or bombs that it finds.

The Buffalo MRAP first entered service in 2003.

A Buffalo MRAP uses its robotic arm to sweep the route ahead of it for mines and explosive devices.

DESIGN AND USE

The bottom of the Buffalo is V-shaped. If an explosive goes off below an MRAP, the shape directs the force away from the people inside. The Buffalo has other features to help keep the crew safe and the vehicle running. It is armored to protect against gunfire. The tires are able to keep the vehicle moving even if they go flat. The Buffalo's engine runs on diesel fuel, but it can also use jet fuel if needed.

BV 206

The Bv 206 is a powerful, flexible transport vehicle. It is articulated, meaning that it is made of two separate parts that are connected but can move independently. This makes the Bv 206 very maneuverable. The vehicle runs on tracks, like a tank. The large tracks spread out the vehicle's weight over a large area. As a result, the Bv 206 puts less pressure on the ground than a person's foot. This makes it great for traveling over snow, sand, and marshes.

The front part of the vehicle can hold up to six soldiers, and the rear part can hold up to 11. The Bv 206 can also be

Members of the Marine Corps install communications equipment onto a Bv 206 that will allow troops on the ground to coordinate movements more effectively.

The Bv 206 is an amphibious vehicle built to handle all kinds of terrain, including steep, snowy areas.

loaded with approximately 5,000 pounds (2,270 kg) of supplies. Different versions of the vehicle have different jobs. Some carry artillery. Others hold anti-tank weapons. The Bv 206 can be used as an ambulance too.

HISTORY

The Bv 206 was designed in Sweden in the 1970s. It entered service with the Swedish military in 1980. It was later sold to the US Army and other military forces around the world.

C-5M SUPER GALAXY

The C-5M Super Galaxy is a large cargo aircraft. It is the Air Force's biggest plane. The original C-5A Galaxy entered service in 1970. Upgrades to the

Members of the Air Force help load a truck through the nose of a C-5M Super Galaxy.

Wheels extend from a Super Galaxy as the aircraft prepares for landing at an air base in California.

C-5M began in the early 2000s. This model's improved engines let it take off from shorter runways and quickly gain altitude.

CARGO CAPACITY

The Super Galaxy has a crew of seven people. This includes a pilot and copilot, two flight engineers, and three loadmasters. The loadmaster's job is to safely secure the cargo so it doesn't move around during flight. The aircraft's nose lifts up so cargo can be loaded from the front. Cargo can also be loaded through a ramp in the back.

The aircraft can carry 281,000 pounds (127,460 kg). That means it is able to carry two M1 Abrams tanks at once. It could also carry 16 trucks or five helicopters. To support all that weight, the plane has five sets of landing gear with 28 wheels in all. The C-5M Super Galaxy is a vital tool for carrying the military's biggest cargo.

C-17 GLOBEMASTER III

The C-17 Globemaster III carries large cargo for the Air Force. It delivers vehicles and supplies to distant bases. This helps the military quickly respond to threats and challenges around the world. The C-17 entered

The C-17 is one of the largest military planes in the world.

Despite its large size, the C-17 is maneuverable and can take off and land on short, narrow runways.

service in 1993, and it has been an important tool for the Air Force ever since.

CARGO AND CREW

The C-17 has a large ramp in the back for loading cargo. The gigantic cargo hold is 88 feet (27 m) long. This allows it to hold some of the Army's largest vehicles, including the M1 Abrams tank. The plane can also deliver paratroopers. It can hold 102 troops along with their parachutes and other equipment. In all, the C-17 can carry 170,900 pounds (77,520 kg) of cargo.

A crew of just three people operates this huge aircraft. Each plane has a pilot, a copilot, and a loadmaster. The small crew size reduces costs and training needs.

C-130J HERCULES

The C-130J Hercules is a cargo plane. It enters dangerous areas near the front lines to deliver troops and equipment. The tough plane can take off and land on rough runways, including dirt airstrips. Air Force pilots fly the C-130J all around the world. Besides carrying military gear, some C-130Js are used to deliver supplies to Antarctica, study the weather, or fight fires.

A Hercules aircraft flown by the US Air Force takes off from an air base in Japan.

The C-130J can reach flying speeds of 417 miles per hour (671 kmh).

The C-130J has a ramp in the back that lets crews load large pieces of cargo. The plane can carry up to 42,000 pounds (19,050 kg). Helicopters and light armored vehicles are among the biggest objects it carries. The plane has a crew of five. This includes two pilots, a navigator, a flight engineer, and a loadmaster.

HISTORY

The first C-130 entered service in 1956. Many more versions have come since then. The C-130J model joined the Air Force in 1999. It flies faster and farther than earlier models.

CH-47F CHINOOK

The CH-47F Chinook is the only Army helicopter designed to lift heavy cargo. It is mainly used to transport troops and their equipment. The helicopter has also been used for medical evacuations, search and rescue missions, and disaster relief.

TECHNOLOGY AND CARGO

The original version of the helicopter, the CH-47A, went into service in 1962. The upgraded CH-47F arrived in 2006.

Nineteen countries use the Chinook in their militaries.

Chinooks can carry vehicles and weapons, such as M777 howitzers.

It features stronger construction and upgraded electronic equipment. This gear gives pilots more information and helps them fly safely. The helicopter has a crew of three, consisting of a pilot, copilot, and flight engineer. Though the CH-47F is not designed for combat, it can be equipped with up to three machine guns for self-defense.

The Chinook can carry up to 36 troops. Cargo is held within the aircraft. The Chinook can also carry things hanging beneath it on a sling. The sling system can hold 28,000 pounds (12,700 kg) of cargo. This lets it deliver supplies and light vehicles anywhere they are needed.

COUGAR MRAP

The Cougar MRAP vehicle is used by the Army and the Marine Corps. Like other MRAPs, it is a vehicle designed to survive explosions. Its V-shaped hull directs the explosive force away from the passengers. Roadside bombs were a threat during the wars in Afghanistan and Iraq. The Cougar and other MRAPs were developed in response.

> The Cougar has been used by the US military since 2004, but other countries have had vehicles of similar design since the 1980s.

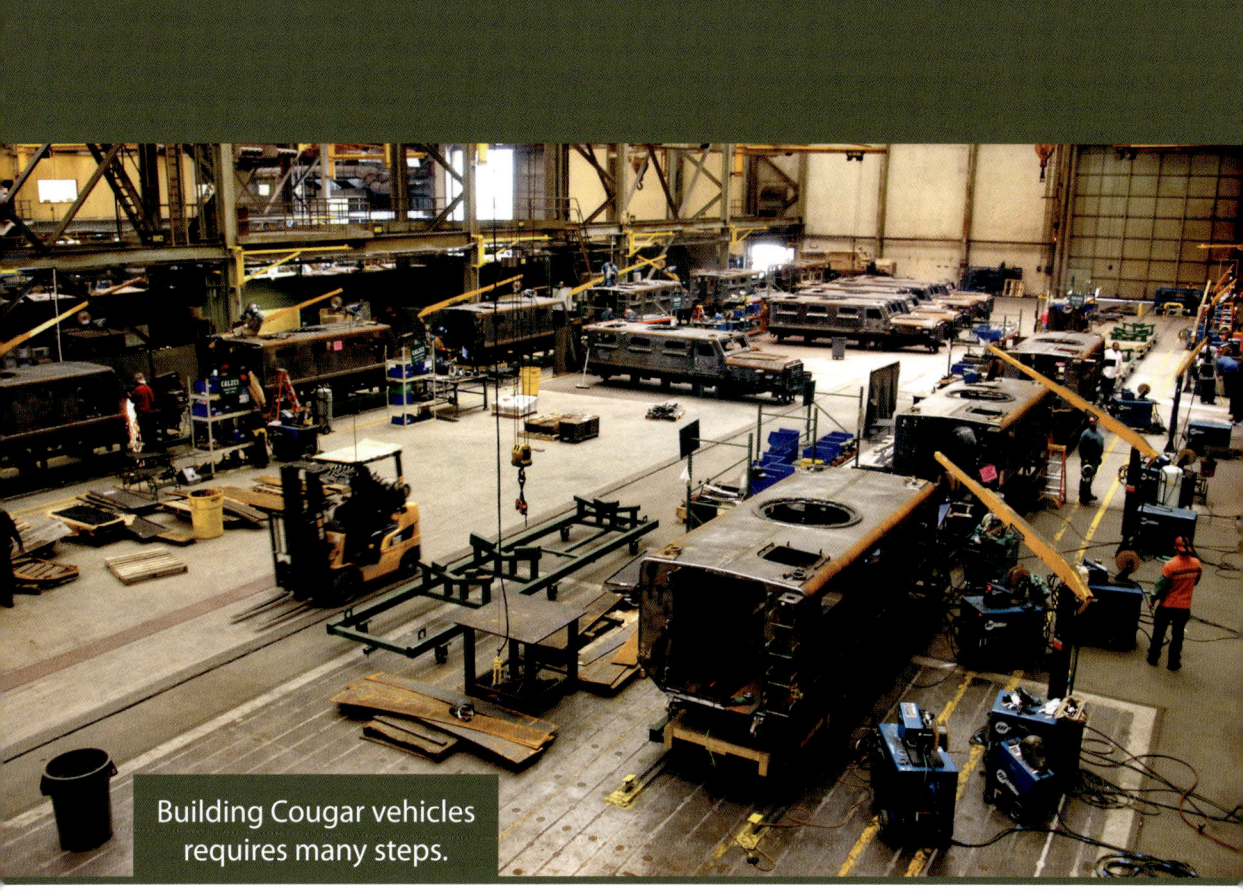

Building Cougar vehicles requires many steps.

DESIGN

Some versions of the Cougar have four wheels, and others have six. Two drivers sit in the front, and up to four passengers can sit in the back. Each seat has safety harnesses, and the vehicle's armor protects against rifle fire. An air filter system protects the crew from poisonous gases outside. If the Cougar needs to escape danger, it can drive away at a top speed of 65 miles per hour (105 kmh).

Crews can mount a machine gun on the top of the vehicle for self-defense. Sometimes a crew member operates the gun directly. Other times, an operator uses a remote-controlled system to aim and shoot safely from inside the Cougar.

E-2 HAWKEYE

The E-2 Hawkeye is an airborne command and control aircraft flown by the Navy. This type of aircraft uses radar to detect vehicles in the air, on the water, and on land. The crew members determine which vehicles are enemies. Then they pass this information to friendly forces.

The most distinctive feature of the E-2 is the huge radar dome attached to the top. The dome is 24 feet (7.3 m) in diameter. The radar system can detect targets more than 340 miles (550 km) away. The Hawkeye operates from aircraft carriers. It helps protect the fleet from enemy aircraft.

The Hawkeye weighs about 26 tons (23.6 metric tons) and flies slower than many other military planes.

The wings of an E-2C Hawkeye can be folded so that the plane takes up less space on an aircraft carrier.

The Hawkeye has a crew of five. A pilot and a copilot fly the plane, and three people operate the radar system. The copilot can also work as a fourth radar operator if needed.

HISTORY

The original E-2 Hawkeye entered service in 1964. Since then, it has seen many upgrades. The Navy began using the E-2D Advanced Hawkeye in 2014. Its radar is even more powerful than the equipment in previous models.

E-3 SENTRY

The E-3 Sentry is an airborne warning and control system (AWACS) aircraft. The Air Force operates it. The Sentry uses powerful radar to give commanders an accurate view of the battlefield. The radar dome it carries is 30 feet (9.1 m) wide. The radar's range is more than 250 miles (400 km). The first E-3 entered service in 1977. It was based on the Boeing 707 jet airliner. Upgrades over the years have improved the radar and added modern computer systems.

IDENTIFICATION SYSTEM

In addition to its radar, the E-3 has an IFF (identification, friend or foe) system. This shows which targets are friendly and which are

An E-3 Sentry gathers information from the air to help inform combat decisions on the ground or at sea.

Crew members must work together to gather surveillance and communicate battle strategy with military leaders.

enemies. Display consoles inside the aircraft relay this information to the crew. The crew includes four people to fly the plane and up to 19 to operate the radar equipment. The data from the E-3 is passed along to military leaders on land or at sea.

E-8 JOINT STARS

The E-8 Joint Surveillance Target Attack Radar System, or Joint STARS, is an Air Force aircraft. It is used for battle management, command and control, and surveillance. Its role is to spot targets on the

In addition to combat and surveillance, the Joint STARS is also used in humanitarian missions.

An E-8's radar antenna is protected in a tube-shaped casing underneath the plane.

ground and send this information to Army and Marine Corps commanders on the ground. This lets the US ground forces plan their missions.

The E-8 is a modified version of the Boeing 707 airliner. To find targets, the crew uses a powerful radar system. The long, thin radar antenna hangs under the front of the plane. It can detect objects over an area of about 19,300 square miles (50,000 sq km). The radar is used mostly for ground targets, but it can also spot helicopters and some slow-moving planes.

CREW

It takes four people to fly the Joint STARS. Other crew members help operate the aircraft's systems. The exact number of crew members varies depending on the mission. A typical mission might have 15 people from the Air Force and three from the Army to operate the radar system.

EA-18G GROWLER

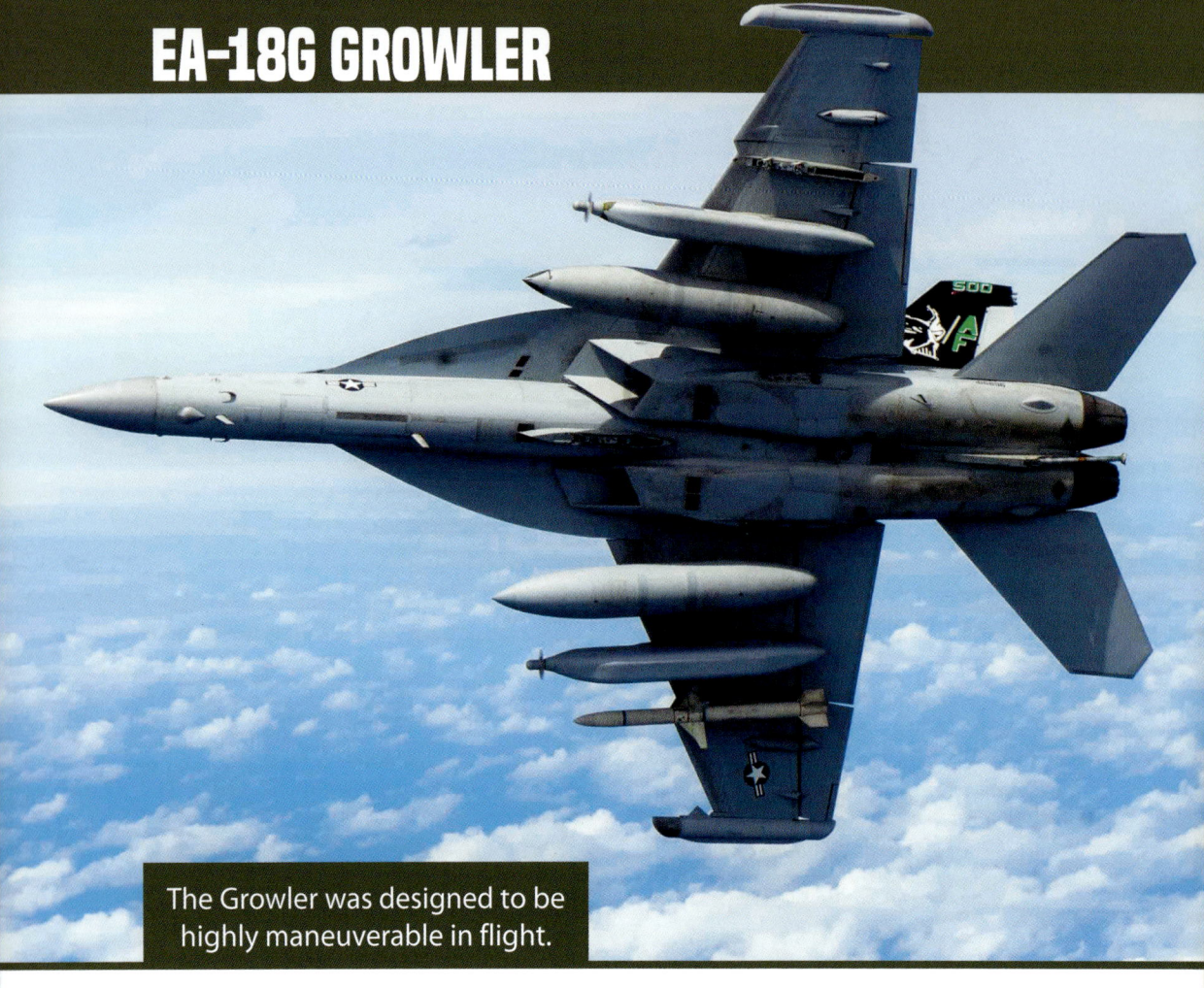

The Growler was designed to be highly maneuverable in flight.

The EA-18G Growler is an electronic warfare aircraft used by the Navy. It is a modified version of the F/A-18F Super Hornet fighter jet. The Growler combines the power, speed, and agility of the Super Hornet with new electronic warfare capabilities.

Electronic warfare involves jamming enemy signals. The Growler has a few systems to do this. ALQ-218 sensors on the wingtips can detect and identify the sources of radio signals. ALQ-99 jamming pods interfere with enemy radar and communications. Jamming prevents enemy aircraft from locating and attacking US aircraft.

CREW AND WEAPONS

The Growler has a crew of two. The pilot sits in the front, and a weapon systems officer sits in the back. Besides its electronic warfare mission, the Growler can defend itself too. It can carry air-to-air missiles to shoot down other planes. It can also use air-to-ground missiles to destroy enemy radar stations.

A Growler lands on the deck of an aircraft carrier.

F-15C EAGLE

The F-15C Eagle is a tactical fighter jet. It is designed mainly for air-to-air combat. The Eagle's mission is to defeat other aircraft in dogfights, which are close-range air battles.

The Eagle's powerful engines give it quick acceleration and a high top speed. Its wing design makes the plane very maneuverable. The fighter can make tight turns without losing much speed. A heads-up display (HUD) shows important information, such as speed and altitude. It is right in front of the pilot. This way, the pilot does not need to look down to get

The F-15 can climb from ground level to an altitude of 65,000 feet (19,810 m) in just over two minutes.

Eagles fly together in a tight formation during an aerial operations training exercise.

this information and can focus on flight. All these parts of the design are important in intense dogfights.

The Eagle can carry eight air-to-air missiles. They can be a mix of AIM-120 and AIM-9 missiles. The Eagle also has a 20-mm cannon that it can use for dogfights and ground attacks.

PAST AND PRESENT

The first F-15s flew in 1972. The F-15C upgrade arrived in 1979. The jet has seen many improvements since then. Modern Eagles have better radar, more advanced computers, and improved missiles.

F-15E STRIKE EAGLE

The F-15E Strike Eagle is based on the F-15C Eagle. The first F-15E entered service with the Air Force in 1988. While the Eagle was mainly for air-to-air combat, the Strike Eagle is primarily used to hit ground targets. The Strike Eagle is designed to fly into enemy territory, drop bombs or missiles, and then fight its way back out.

In 2019, the Air Force had 219 Strike Eagles.

As engines have improved, the Strike Eagle has seen upgrades that have resulted in quicker takeoff times.

The aircraft has a crew of two, with a pilot and a weapon systems officer (WSO). The crew uses an advanced radar system to detect enemies in the air and on the ground. The pilot keeps track of enemy aircraft while the WSO plans an attack on ground targets.

TECHNOLOGY

The Strike Eagle uses the Low Altitude Navigation and Targeting Infrared for Night, or LANTIRN, system to carry out its missions. The system is made up of two pods. One contains radar that senses the terrain, letting the pilot fly safely at low altitudes. The other contains a laser targeting system to accurately hit targets.

F-16C FIGHTING FALCON

The Fighting Falcon can fly at a top speed of 1,500 miles per hour (2,410 kmh).

The F-16 is a single-engine fighter jet used by the Air Force. Its official name is the Fighting Falcon, but many pilots prefer the unofficial nickname Viper. The F-16 was designed to be a lighter, cheaper airplane than the F-15 Eagle. It can fire air-to-air missiles and drop air-to-ground weapons. It also carries a 20-mm cannon.

The F-16 originally entered service in 1979. Today's models are the F-16C and F-16D. The C model is the more common single-seat version. The D has two seats and is usually used for training.

DESIGN

The Fighting Falcon is a highly maneuverable aircraft. It can make sharp turns at high speeds. This generates harsh forces

on the pilot. To help the pilot withstand this, the seat is angled back farther than in most planes. The control stick is mounted on the side of the cockpit rather than in the center. This makes it easier to control the jet while doing intense maneuvers.

A member of the Air Force inspects a missile that is attached to a Fighting Falcon.

F-22A RAPTOR

The F-22A Raptor is a stealth fighter jet. Its stealth features help the aircraft hide from enemy radar systems. Radar systems send out waves of energy. They then detect the reflections of these waves to locate objects. The Raptor's shape and materials prevent radar waves from reflecting back.

The Raptor began as the YF-22. It was competing against another plane, the YF-23, to be the Air Force's next-generation fighter jet. The Air Force chose the YF-22 in 1991. After many more years of development and testing, it entered service as the F-22A Raptor in 2005.

The Raptor's aerodynamic design allows it to supercruise. This means it can fly at high speeds without using much fuel.

Each F-22A is crewed by a single pilot.

DESIGN AND WEAPONS

The F-22A is a fast, maneuverable jet. Its two powerful engines can use thrust vectoring. This means they can angle upward or downward instead of just straight back. This makes the plane highly maneuverable.

A Raptor usually carries six AIM-120 and two AIM-9 air-to-air missiles. Unlike most fighters, it carries these weapons in its belly rather than on its wings. A door on the bottom of the plane opens, the missile fires, and the door closes again. Maintaining a smooth exterior helps keep the Raptor stealthy.

F-35 LIGHTNING II

The Lightning II's powerful engine allows the aircraft to fly at high speeds.

The F-35 Lightning II is an advanced fighter aircraft. This stealth fighter features high-tech radar, sensors, and targeting systems. Its pilot wears a helmet-mounted display to show key information at all times. The plane also has data links to share this information with other planes and friendly military forces.

DIFFERENT VERSIONS

There are three different versions of the F-35, each used by a different military service. The Air Force flies the F-35A. This model takes off and lands on normal runways. The Marine Corps uses the F-35B. This version uses a device called a lift fan to help it take off and land in short distances or even vertically, allowing it to operate from small bases. The Navy flies the F-35C. This model is designed to operate on aircraft carriers.

The three versions of the F-35 have much in common, but there are key differences. The F-35A and B are similar, except that the B model's lift fan takes up space that the A model uses for fuel. This gives the F-35B a shorter range. The F-35C has the strongest structure so it can handle rough carrier landings. It also has the largest wings, helping it take off more easily from carriers.

F-35 aircraft can be refueled in the air, giving these planes an unlimited range of service.

F/A-18 HORNET

The F/A-18 Hornet is a fighter jet used by the Navy and the Marine Corps. It operates from aircraft carriers or bases on land. The *F* in the Hornet's name stands for "fighter" and the *A* for "attack." There are two models. The C model is a single-seat aircraft. The D model has two seats. The pilot sits in the front, and the weapon systems officer is in the back.

A Hornet fighter jet prepares to land on the deck of an aircraft carrier.

The Hornet has two powerful engines that can push the jet to supersonic speeds.

The Hornet is a capable fighter. It carries an M61 cannon and air-to-air missiles. It is also good at attacking ground targets with bombs and air-to-surface missiles. This flexibility gives commanders many options when planning missions.

HISTORY AND USE

The F/A-18A entered service in 1983. The C and D models arrived in 1987. The newer jets were able to fire the latest air-to-air missiles. They also added equipment to prevent enemy missiles from working properly.

F/A-18 SUPER HORNET

The F/A-18 Super Hornet looks similar to the F/A-18 Hornet C and D models. However, key design changes differentiate these models. The Super Hornet is larger than the Hornet. It has more powerful weapons and carries more fuel. It can also hold more weapons and has better electronics.

The experienced Blue Angels pilots show off the agility of the Super Hornet.

A Super Hornet, *top*, refuels an EA-6B Prowler, *bottom*, while both aircraft are in flight.

Like the Hornet, the Super Hornet flies a mix of air-to-air and air-to-ground missions. There are two models of Super Hornet. The E model has one seat, while the F model has two. The F/A-18F is used for training and for missions that require a second crew member.

HISTORY AND USE

The Super Hornet entered service with the Navy in 1999. Over the years, it has seen upgrades that are known as Blocks. Block II added a new radar system. Block III added touchscreens and more powerful computers.

The Super Hornet is not used only for combat. The Navy's Blue Angels flight demonstration team flies the Super Hornet. The team's pilots skillfully fly in formation in front of crowds at airshows. Their job is to promote the Navy and help with recruitment.

FAMILY OF MEDIUM TACTICAL VEHICLES (FMTV)

FMTV are heavily armored to protect troops and supplies.

The Family of Medium Tactical Vehicles (FMTV) is a set of trucks used by the Army. They have a wide variety of jobs. Some deliver ammunition and other supplies. Others help recover damaged vehicles. Some types of FMTV have weapons mounted on them. The FMTV is designed to be highly reliable. All FMTV vehicles share common parts. This makes repairs easy and efficient. Mechanics do not need specialized parts if something in the vehicle breaks.

DESIGN

There are two basic types of FMTV. The Light Medium Tactical Vehicle can carry 2.5 tons (2.3 metric tons) and tow 6 tons (5.4 metric tons). The Medium Tactical Vehicle carries

5 tons (4.5 metric tons) and tows 10.5 tons (9.5 metric tons). Either type can be hooked to different trailers depending on the mission.

The Medium Tactical Vehicle can also have the High Mobility Artillery Rocket System (HIMARS) installed on it. HIMARS is a long-range rocket launcher. After firing, the crew can quickly drive away before the enemy fires back.

Soldiers unload supplies from the back of a Light Medium Tactical Vehicle to assist Texas residents affected by a hurricane in 2008.

FAMOUS-CLASS CUTTER

The Famous-class cutter is one of two main types of Medium Endurance Cutters used by the Coast Guard. These ships are used for law enforcement and search and rescue missions. Each ship has a crew of 100 people. Its weapons include one Mk-75 naval gun and two machine guns.

Famous-class cutters are 270 feet (82 m) long.

Crew members of the USCGC *Mohawk*, a Famous-class cutter, pose next to an H-65 Dolphin.

To help with search and rescue, the Famous class can support the H-65 Dolphin helicopter. The ship has a system to help it remain stable in rough waters. This makes it safer for helicopters to take off during storms. The ships usually carry a Dolphin and five aviation crew members when going out to sea. The cutter can also carry small boats that are used to board vessels during law enforcement missions.

ADVANCEMENTS OVER THE YEARS

The Famous class entered service in 1983. The first ship was the USCGC *Bear*. The class has since received many updates. It saw computer upgrades in the 1990s, followed by another round of upgrades in the 2010s.

FREEDOM-CLASS LITTORAL COMBAT SHIP (LCS)

The Freedom-class littoral combat ship (LCS) is one of two classes of LCS used by the Navy. These vessels are designed to fight in the littoral zone, meaning the area near the shore. Threats in the littoral zone include mines, submarines, and small, fast surface craft. The LCS can operate in the open ocean if needed, but its main focus is on the coastline.

Freedom-class ships are designed to be fast and maneuverable. They have a shallow draft, meaning the ship doesn't extend far below the surface. This makes it easier to

Over the years, Freedom-class littoral combat ships have received many upgrades that have increased their effectiveness in conflict, such as a greater targeting range.

Helicopters can land on and take off from Freedom-class littoral combat ships.

sail in shallow coastal areas. The ship's advanced automation technology reduces the number of people needed to operate the ship. The crew size is just 50 people.

SHIP DESIGN

The ship's weapons include one Mk-110 naval gun and various surface-to-air missiles. The rear of the Freedom-class ship features a flight deck larger than those on most Navy ships. Next to it, there is a hangar with space for two helicopters.

GERALD R. FORD–CLASS AIRCRAFT CARRIER

The Gerald R. Ford–class aircraft carrier is named for former president Gerald R. Ford, who served in the US Naval Reserve during World War II. It is the newest type of aircraft carrier used by the US Navy. The first ship in the class, the USS *Gerald R. Ford*, entered service in 2017.

ADVANCED TECHNOLOGY

The Ford class has many important innovations. The first has to do with power. US carriers use nuclear reactors to generate electricity. This power propels the ship and operates equipment. The Ford's two reactors are much more powerful than those used in older types of aircraft carriers. Yet advancing technology also means the reactors are smaller and simpler than in the past, making maintenance easier.

When launched, the USS *Gerald R. Ford* was the largest aircraft carrier in the world. It was more than 1,090 feet (332 m) long.

The USS *Gerald R. Ford* can carry more than 75 aircraft at once.

Older carriers launch aircraft using steam-powered catapults. The Ford class does this with electricity instead. It uses the electromagnetic aircraft launch system (EMALS). Compared with steam catapults, EMALS is smaller, more efficient, and easier to control. It can launch both large and small aircraft, including drones.

GLOBAL POSITIONING SYSTEM (GPS) SATELLITES

It takes about 12 hours for a GPS satellite to orbit Earth.

In 2023, there were about 30 active Global Positioning System (GPS) satellites. There are multiple versions in orbit. GPS satellites enable the location-finding functions in smartphones and other devices. The Space Force controls these satellites from Schriever Space Force Base in Colorado.

GPS was originally developed for military use. It is useful for navigating land vehicles, ships, and planes. It can be used to precisely target weapons too. But GPS is also helpful in civilian life. People use it while driving, hiking, and more.

HOW GPS WORKS

Each GPS satellite sends out signals. Devices on the ground receive those signals. By receiving signals from multiple satellites and comparing their timing, devices can figure out their exact location on Earth.

As technology has advanced, the reliability and accuracy of GPS has improved. New versions of the satellites have made this possible. The first GPS IIIF satellite launched in 2018. These satellites are designed for a 15-year life span. They are powered by solar panels, and each weighs about 5,000 pounds (2,270 kg).

GPS satellites are designed so that their solar panels continuously orient themselves toward sunlight.

HC-144 OCEAN SENTRY

The HC-144 Ocean Sentry is a surveillance airplane flown by the Coast Guard. It has many uses. During law enforcement missions, it searches for people transporting illegal drugs at sea. When responding to natural disasters, it flies over land to assess the damage. In search and rescue missions, it locates victims of shipwrecks or plane crashes. In each case, the Ocean Sentry is used to locate things and then pass that information along to other responders.

To accomplish these missions, the airplane carries enough fuel to fly for ten hours at a time. Its advanced sensors scan the land and sea below. The Ocean Sentry can also drop supplies such as life rafts and flares to people in the water.

The Coast Guard uses seven types of aircraft, including the HC-144 Ocean Sentry.

The Ocean Sentry can carry up to 40 passengers.

MODERN UPGRADES

In the 2020s, the Coast Guard began upgrading its Ocean Sentry aircraft with a system called Minotaur. This system adds improved sensors, radar, and computers. Once the upgrade is made, the aircraft is known as the HC-144B.

HEALY-CLASS ICEBREAKER

The *Healy* is a Coast Guard icebreaker. These types of ships are built with strong hulls to clear a channel through thick sea ice in cold regions. The *Healy* is one of the largest Coast Guard ships, at 420 feet (128 m) long. It can break through 4.5 feet (1.4 m) of ice while moving at a speed of 3 knots (5.6 kmh). The ship can operate in temperatures of −50 degrees Fahrenheit (−46°C).

The USCGC *Healy* is able to break through thick ice in the Arctic Ocean.

USE

The *Healy* was designed mainly for scientific research in the Arctic region. Scientists helped plan its labs and equipment while the ship was being designed and constructed. It includes living space for up to 50 scientists. The *Healy* can also carry two helicopters. In 2015, the *Healy* made the first solo trip to the

North Pole by a US surface ship. It repeated the journey in 2022.

The ship has another mission besides research. It gives the US military a presence in the Arctic region. The federal government sees this as a strategically important duty.

Ice breaking allows smaller boats to travel through frozen waters.

HEAVY EXPANDED MOBILITY TACTICAL TRUCK (HEMTT)

The Heavy Expanded Mobility Tactical Truck (HEMTT) is an Army vehicle. This large, eight-wheeled truck entered service in 1985. It has a crew of two, and its range is about 300 miles (480 km). The truck can climb steep slopes and drive through water that is 4 feet (1.2 m) deep. Troops have nicknamed it the Dragon Wagon.

MISSIONS AND CARGO

The most common mission of the HEMTT is to resupply vehicles and weapons. But there are many versions of the HEMTT. Each version has a different job and a unique name made up of the letter *M* and a number.

Some versions of the HEMTT are used to tow other large vehicles.

Soldiers load guided rocket ammunition onto the bed of an HEMTT.

The M977 and M985 carry ammunition and other equipment. The M985 also carries guided missiles. The M978 can haul up to 2,500 gallons (9,460 L) of fuel to help keep other vehicles moving. The M984 uses cranes and winches to help recover stuck or broken vehicles. The M1977 carries equipment that lets soldiers quickly build bridges over rivers.

HUMVEE

Most Humvees can seat four people, but the number varies slightly depending on the version of the Humvee.

The US military's High Mobility Multipurpose Wheeled Vehicle (HMMWV) is often known as the Humvee for short. It is a lightweight four-wheel-drive truck. The Humvee is used by the Army, the Marine Corps, the Navy, the Air Force, and the Coast Guard. The first one entered service in 1985. Since then, more than 280,000 have been built.

There are more than a dozen versions of the Humvee. Some carry cargo or troops, some haul weapons, and some are ambulances. Each version shares the same engine and transmission. This makes maintenance and repairs easier and cheaper.

Humvees can be transported by cargo planes. Three fit into a C-130 Hercules, and 15 fit into a C-5 Galaxy. The vehicle can even be airdropped using a parachute. Helicopters can also carry a Humvee using a sling system.

DESIGN

Humvees are designed to tackle tough terrain. They can climb steep slopes. Their wide design makes it hard for them to roll over. Some models feature a tire inflation system. This lets the driver change the tire pressure to get a better grip on the terrain.

Humvees can maneuver through uneven, muddy terrain.

HUSKY MOUNTED DETECTION SYSTEM (HMDS)

The Husky Mounted Detection System (HMDS) combines a vehicle called the Husky with equipment that detects underground explosives. The HMDS locates and marks these threats. This helps clear the way for other vehicles in dangerous areas.

The Husky vehicle was designed in South Africa in the 1970s. Its V-shaped hull directs blast forces away from the crew inside. It has bulletproof windows for additional protection. The Husky has a crew of two: a driver and a sensor operator. Today, the vehicle is in service with the US Army and the US Marine Corps. It was used on missions in the Afghanistan War.

MINE DETECTION

The HMDS uses ground-penetrating radar to detect mines and improvised explosive devices (IEDs). It emits radio waves into the ground and then senses reflections. The vehicle's sensors can tell if an underground object is likely to be an explosive. Some Husky vehicles also have metal detectors, increasing their ability to detect underground objects.

Soldiers drive Huskies during a training exercise to detect mines and other explosives.

The detection equipment is mounted to the front of a Husky vehicle.

INDEPENDENCE-CLASS LCS

The Independence-class LCS is one of two classes of LCS used by the Navy. Both types of ship are designed for missions near coastlines. The most distinctive feature of the Independence class is its trimaran design. This means it has three separate hulls—one center hull and two side hulls. This design gives the ship more stability at sea.

The trimaran design was originally developed for a cruise ship. The increased stability would improve passenger comfort. The concept later

The USS *Independence* and the USS *Coronado* are two of the Navy's Independence-class littoral combat ships.

The Independence class was designed to be speedy, agile, and able to perform many kinds of missions.

evolved into the Independence class, which the Navy approved in 2003. The first ship entered service in 2010.

WEAPONS AND EQUIPMENT

Ships in the Independence class use a modular design. This means the crew can swap out equipment and weapons depending on the mission. Different sets of gear are used for sinking submarines, detecting mines, and fighting ships. To help with its missions, the ship can carry two SH-60 Seahawk helicopters or one CH-53 Sea Stallion helicopter. It can also carry multiple drones.

JOINT LIGHT TACTICAL VEHICLE (JLTV)

The JLTV is designed to travel off-road and across all types of terrain.

The Joint Light Tactical Vehicle (JLTV) is designed to replace existing light vehicles, such as Humvees. The Army and Marine Corps worked together to develop the new vehicle. It comes in two versions. The four-seat Combat Tactical Vehicle is designed for fighting. The two-seat Combat Support Vehicle carries supplies.

DESIGN AND USE

The JLTV has more protection and payload compared with the Humvee. It has more mobility and agility than MRAP vehicles.

The vehicle includes modern technologies. Data links to other vehicles give troops more information on their surroundings and give commanders a better view of the battlefield. The JLTV is designed to be upgraded over time. One possible addition is batteries that would power equipment even when the engine is not running. This would save fuel.

The JLTV entered service in 2019. It is used by the Army, the Marine Corps, the Air Force, and the Navy. It has also been exported to other countries.

Because of its light weight, the JLTV can be carried by helicopters such as the CH-53E Super Stallion.

JUNIPER-CLASS BUOY TENDER

The Juniper-class buoy tenders of the Coast Guard are responsible for maintaining buoys and other structures at sea. Buoys are placed at ports and in waterways to help ships safely navigate. Juniper-class cutters make sure the buoys are functioning correctly. They have a large crane to lift buoys onto the deck for servicing.

The cutters must accurately control their position to work on buoys. They use their engines and sensors to do this. They are able to remain within a 33-foot (10 m) circle even when winds are blowing at 30 knots (56 kmh) and waves are reaching 8 feet (2.4 m) high.

Some buoy tenders work inland or on lakes, but Juniper-class cutters are seagoing vessels.

Crew members perform maintenance on a buoy on the deck of a Juniper-class cutter.

CREW

Juniper-class cutters have a crew of about 45 people. Besides the crew's buoy-tending work, it also performs law enforcement and search and rescue missions. The crew may operate the cutter's two machine guns if needed. The first Juniper-class cutter went into service in 1995. The class replaced buoy tenders that dated back to World War II.

KC-10 EXTENDER

The KC-10 Extender is a tanker and cargo aircraft used by the Air Force. Its main mission is aerial refueling. It links up with other planes in midair and transfers fuel into their tanks. This boosts the range of the receiving planes, making longer missions possible.

The plane is a modified version of the DC-10 airliner. The Air Force began using the KC-10 in 1981. The aircraft contains six large fuel tanks. In all, it can carry more than 356,000 pounds (161,480 kg) of fuel.

A KC-10 Extender refuels an F-16 Fighting Falcon.

In addition to aerial refueling, the KC-10 is used to carry troops and equipment.

REFUELING METHODS

The KC-10 Extender has two systems for refueling planes. The first system is called a boom. This rigid tube extends downward from the back of the plane. An operator controls it, viewing the receiving plane from a window. The boom connects to the receiving plane and transfers fuel at up to 1,100 gallons (4,160 L) per minute. The second method is a hose and drogue system. A flexible hose extends from the plane. It has a funnel-shaped part called a drogue at the end. The drogue connects to the receiving plane, transferring up to 470 gallons (1,780 L) of fuel per minute.

KC-46A PEGASUS

The KC-46A Pegasus is an aerial refueling plane. The first Pegasus was delivered to the Air Force in 2019. The jet is a modified version of the Boeing 767 airliner. It carries more fuel and cargo than the older KC-135.

A member of the Air Force inspects a KC-46A to make sure its engines are operating properly.

A KC-46A Pegasus extends a refueling boom to an F-15E Strike Eagle.

REFUELING TECHNIQUES

Like earlier air tankers, the Pegasus can use either a boom system or a hose and drogue system to refuel planes. Refueling pods on the wings let it transfer fuel to multiple airplanes at the same time. But unlike with past planes, the boom operator does not sit in the back of the plane and look out a window. Instead, the boom operator sits in the front with the rest of the crew. He or she uses cameras, sensors, and screens to handle refueling. One pair of cameras even captures 3D video. The operator wears special glasses to see a 3D view of the refueling. This makes it easier to judge depth and distance.

KC-135 STRATOTANKER

The Air Force has about 400 Stratotankers.

The KC-135 Stratotanker is an Air Force aerial refueling plane. It helps military aircraft fly extended missions around the globe. It can also carry cargo. For medical evacuation missions, it can carry patients in need of medical care.

The KC-135 is based on the Boeing 367-80, an early prototype for a jet-powered airliner. This design led to both the Boeing 707 airliner and the KC-135. The Stratotanker entered service in 1957. Since then, it has received many upgrades. Newer engines have given the plane a greater carrying capacity. These engines are also quieter and more efficient than the old ones.

REFUELING AND CREW

The Stratotanker's main refueling method is a boom system. It can also use a hose and drogue system. Refueling pods on the wings allow it to refuel two planes at once. The plane's three-person crew consists of a pilot, a copilot, and a boom operator. Some missions may have a navigator too. For medical missions, the crew includes nurses and medical technicians.

An A-10C Thunderbolt II receives fuel from the boom system of a Stratotanker.

KEEPER-CLASS BUOY TENDER

The USCGC *Henry Blake*, a Keeper-class buoy tender, is named after a keeper of the New Dungeness Lighthouse in Washington State.

The Coast Guard's Keeper-class buoy tenders are used to place and maintain buoys and other navigational aids. Rather than using a normal propeller and rudder for movement and steering, the Keeper-class cutters use two Z-drive propulsion units. These systems can each turn 360 degrees, allowing the cutter to move precisely in any direction. They are operated by a thruster in the cutter's bow.

The cutter also has advanced computers and navigation systems. It uses a system called differential GPS, which enhances the accuracy of a GPS receiver. This helps the crew accurately place navigation aids. Keeper-class cutters have a crew of 24 people.

KEEPER-CLASS ORIGINS

The Keeper-class buoy tenders are named after lighthouse keepers of the past. They include the USCGC *Ida Lewis* and the

USCGC *Frank Drew*. Each cutter has its own area of responsibility. For example, the *Ida Lewis* works off the coast between New York and Massachusetts. Its crew manages more than 350 buoys and other navigational aids.

Crew members prepare for a buoy to be lifted onto the deck of a Keeper-class buoy tender.

LIGHT ARMORED VEHICLE (LAV-25)

The Light Armored Vehicle (LAV-25) is used by the Marine Corps. The eight-wheeled vehicle was introduced in 1983. It replaced armored personnel carriers that ran on tracks. It is designed to be quicker and more maneuverable than those earlier vehicles. The Marines use the vehicle for reconnaissance and scouting missions, where its speed is especially useful.

Marines drive LAV-25s during some field training exercises.

Marines fire the chain gun of their LAV-25 as part of a live-fire exercise while aboard an amphibious assault ship.

WEAPONS AND USE

The LAV-25's weapons include a powerful M242 chain gun in a turret on top. It also has two M240 machine guns. The vehicle has a crew of three and can hold six troops. When the troops leave the vehicle to scout nearby, the LAV-25 supports them with its heavy firepower. It also has eight launchers for smoke grenades, helping the vehicle and its troops hide from enemy fire.

The LAV-25 is amphibious, meaning it can also function in water. It can drive through streams and rivers. It takes about three minutes to prepare this vehicle for operation in the water.

LEGEND-CLASS CUTTER

The Legend-class cutters are some of the most advanced cutters in the Coast Guard fleet. The first, the USCGC *Bertholf*, entered service in 2008. Legend-class cutters are also the largest combat cutters used by the Coast Guard. Each vessel is 418 feet (127 m) long and has a crew of about 120 people.

WEAPONS AND TECHNOLOGY

The cutter's offensive weapons include a Mk-110 naval gun and four M2 machine guns. It can also support two H-65 Dolphin helicopters. For defense, the cutter has a Mk-53 decoy

Legend-class cutters can be used in many types of roles, including homeland security, law enforcement, and environmental protection.

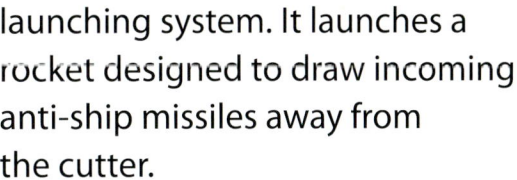
The USCGC *Bertholf* pulls into a pier at the naval base in San Diego, California.

launching system. It launches a rocket designed to draw incoming anti-ship missiles away from the cutter.

The Legend-class cutters have systems for command, control, communications, computers, intelligence, surveillance, and reconnaissance. The military calls this set of abilities C4ISR for short. C4ISR ensures that the cutters can analyze their surroundings and carry out their missions successfully.

LEWIS B. PULLER–CLASS EXPEDITIONARY SEA BASE (ESB)

In 2023, the USS *Lewis B. Puller* was stationed off the coast of Sudan.

The Lewis B. Puller–class Expeditionary Sea Base (ESB) is a large ship that acts as a floating base for helicopters and small boats. The ship's helicopters carry out anti-mine missions, work with special forces troops, and more. Based on an existing design for oil tankers, the ship is 785 feet (239 m) long. Its huge flight deck has room for four helicopters.

The first ship in the class, the USS *Lewis B. Puller*, joined the Navy in 2017. The class is named for Lewis Burwell "Chesty" Puller. Puller received more awards than any other Marine in history. He served in several conflicts, including World War II and the Korean War (1950–1953).

CREW

Some of the crew members on the ship are civil service mariners, known as CIVMARs. They are responsible for navigating, operating the ship's cranes, and maintaining the ship's mechanical systems. A group of about 40 CIVMARs works alongside a military crew of about 100 people.

An MH-53 Sea Dragon helicopter lands on the flight deck of the USS *Lewis B. Puller*.

LOGISTICS VEHICLE SYSTEM REPLACEMENT (LVSR)

The Logistics Vehicle System Replacement (LVSR) is a heavy truck used by the Marine Corps. It runs on ten wheels and can carry cargo on or off roads. It delivers ammunition, food, water, heavy equipment, and other supplies. The LVSR entered service in 2009.

CARGO AND CREW

There are three versions of the LVSR. The MKR 18 carries up to 22.5 tons (20.4 metric tons) of cargo. The MKR 15 is designed to go off-road to retrieve stuck vehicles. It uses a powerful winch to pull vehicles out of water, mud, or other challenging spots. It can rescue Humvees, FMTVs, MRAPs, and more. Finally, the

An LVSR drives to shore after being transported by ship.

The LVSR MKR 16 is designed to carry heavy equipment, including combat vehicles.

MKR 16 is made for towing. It pulls cargo trailers, heavy equipment, or combat vehicles.

The cabin of the LVSR fits two crew members. Armor in the floor of the cabin helps defend against mine explosions. Extra armor and a stronger windshield can be added to give more protection against bombs and gunfire.

LOS ANGELES–CLASS ATTACK SUBMARINE

Los Angeles–class attack submarines are the most common submarines in the Navy. They are designed to attack enemy vessels, including other submarines. Some submarines in this class have Vertical Launching System tubes. These can be used to launch Tomahawk cruise missiles at targets on land or at sea.

Navy sailors crewing the Los Angeles–class attack submarine USS *Olympia* enjoy a swim call, a time to swim freely in the open waters.

Los Angeles–class attack submarines improved upon a 1960s design. Modern submarines have engines that are powerful yet quiet.

The Los Angeles class entered service in 1976. Most of the boats are named for American cities, such as the USS *Chicago* and the USS *Boise*. They are powered by nuclear reactors. The reactor, along with other equipment, is located at the rear of the submarine. The crew, weapons, and control systems are in the front. Los Angeles–class submarines have a crew of about 130 people.

ROLE

Los Angeles–class submarines use four torpedo tubes to fire Mk-48 torpedoes. In addition to fighting ships, they also perform other jobs. They can deliver special forces troops to enemy shores, place sea mines, do reconnaissance, and defend friendly ships.

M1 ABRAMS

The M1 Abrams is the only tank in service with the Army. It is named for General Creighton Abrams, who commanded US troops in the Vietnam War. The original version of the tank was first produced in 1980.

A newer version, the M1A1, entered service in 1986. It added stronger armor and a more powerful M256 main gun. This was followed by the M1A2 Abrams in 1992. This version added armor made from a very dense metal called depleted uranium. Today's models are still called M1A2s, but further upgrades have added better computers, a remote-controlled machine gun, and other features.

An M1A2 Abrams fires its main gun during a training exercise.

M1 Abrams tanks can reach speeds of 42 miles per hour (68 kmh).

WORKING TOGETHER

The Abrams has a crew of four, including a commander, a gunner, a loader, and a driver. The loader reloads the M256 gun by hand. Some other countries use tanks with automatic loading systems, but the US Army found manual loading to be more reliable. The tank can accurately fire at targets more than 2.5 miles (4 km) away.

M2 BRADLEY

The M2 Bradley is crewed by three people: a commander, a gunner, and a driver.

The M2 Bradley is an infantry fighting vehicle used by the Army. Infantry fighting vehicles are designed to safely transport soldiers into combat. The Bradley's armor offers protection against gunfire. It also includes reactive armor blocks. These are pieces of armor containing explosives. They explode outward when hit by high-powered weapons, absorbing some of the weapons' energy. This reduces the damage those weapons can do to the vehicle.

At the same time, the Bradley offers significant firepower of its own. It has a turret with a 25-mm Bushmaster chain gun that can shoot armor-piercing and high-explosive rounds. It also has an anti-tank missile launcher as well as a machine gun.

HISTORY TO TODAY

The original M2 Bradley entered service in 1981. It has been upgraded many times since then. The M2A4 arrived in 2020. It added a new engine and better electronic systems. A related vehicle, the M3 Bradley, is used for armored reconnaissance. The M1283, a version of the Bradley without a turret, is an armored personnel carrier.

The M2 Bradley drives on tracks like a tank.

M9 ARMORED COMBAT EARTHMOVER (ACE)

The M9 Armored Combat Earthmover (ACE) is used by the Army and the Marine Corps. It entered service in 1986. This engineering vehicle supports the work of troops and other vehicles on the ground. It bulldozes, hauls, tows, and winches. The ACE can clear enemy obstacles and dig defensive positions for tanks and artillery. It can excavate trenches to block the movement of enemy tanks. It can also maintain and repair roads, keeping other vehicles moving.

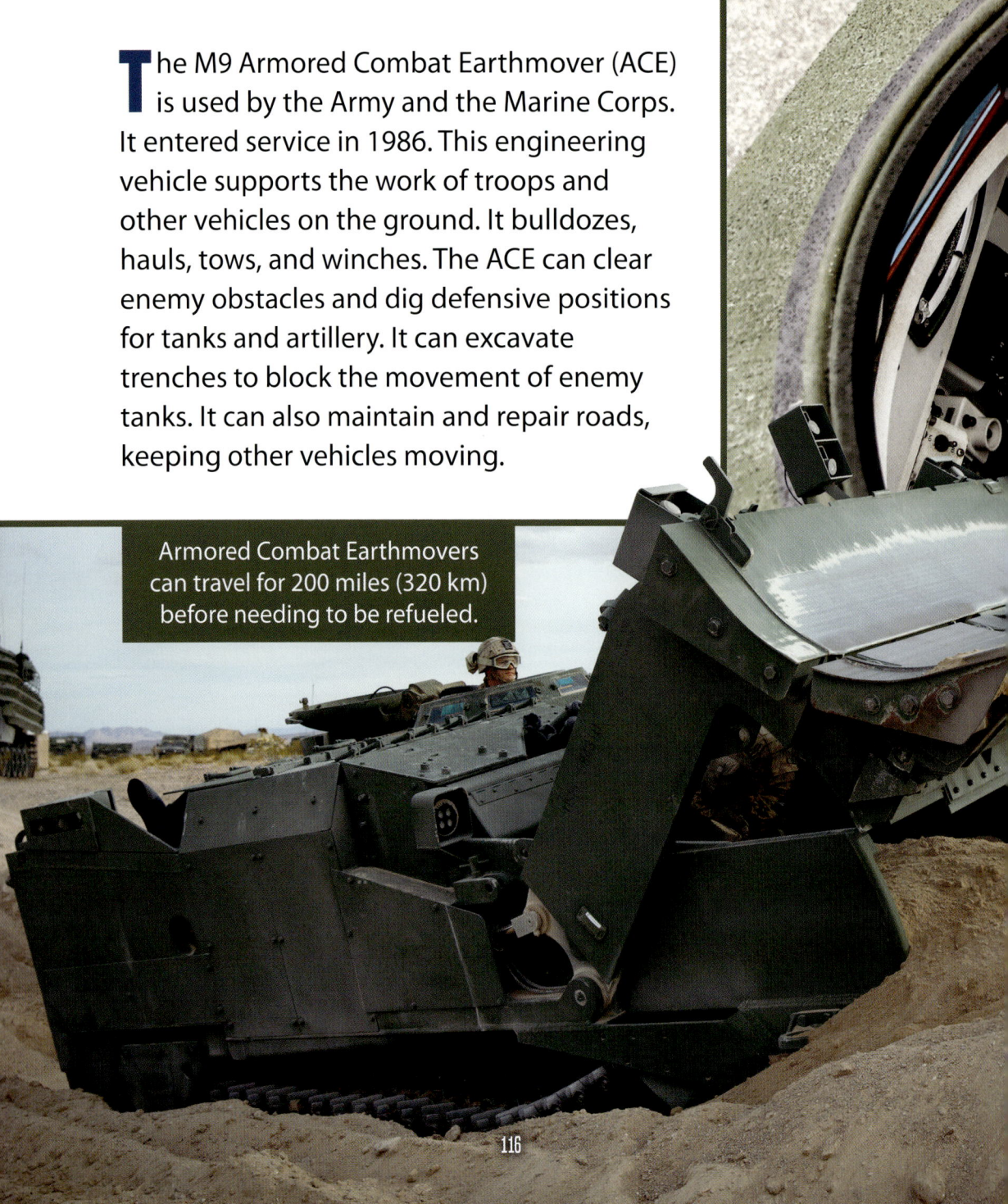

Armored Combat Earthmovers can travel for 200 miles (320 km) before needing to be refueled.

The M9's driver's seat features a variety of joysticks, buttons, and displays.

CREW AND DEFENSE

The vehicle has a crew of one engineer. He or she can raise or lower the front of the vehicle, depending on what is needed for the job. For example, the engineer may need to lower the front when using a bulldozer scoop to dig. Armor protects the engineer from gunfire. The ACE does not carry any weapons. Instead, it is usually joined by an M2 Bradley infantry fighting vehicle for protection. This allows the M9 to work under fire.

M88A2 HERCULES

The M88A2 HERCULES is used by the Army to recover damaged or broken-down vehicles in combat. The name HERCULES stands for Heavy Equipment Recovery Combat Utility Lift and Evacuation System. The armored vehicle protects crews as they bring vehicles back behind the front lines for repairs.

An M88 Recovery Vehicle tows an M1A1 Abrams tank.

Marines prepare an M88A2 HERCULES to recover a vehicle that has become stuck in mud.

The HERCULES uses a powerful winch to recover the Army's heaviest vehicles, including the M1A2 Abrams. It has a crew of three: a commander, a winch operator, and a mechanic. The crew can use a machine gun for self-defense. The vehicle also has smoke generators to hide it from enemy view.

HISTORY AND LOOKING AHEAD

The original M88 entered service in 1961. The upgraded M88A1 arrived in 1977. It was designed to recover the M60 Patton tank that was in service at the time. The M88A2 was introduced in 1997. It added more power for winching, lifting, and towing, allowing it to recover the much heavier Abrams. Anticipated plans upgrade the vehicle to an M88A3 model.

M104 WOLVERINE ARMORED BRIDGELAYER

The M104 Wolverine armored bridgelayer has a simple job. It sets up portable bridges so that other vehicles can cross bodies of water or other obstacles. The Army first used the Wolverine in 2003.

The Wolverine is based on the M1A2 Abrams tank. The bridge equipment sits where the tank's turret would be. The bridge is 13 feet (4 m) wide and 85 feet (26 m) long. It is made of aluminum, and it is strong enough for the Army's heaviest vehicles, such as the

An M104 Wolverine sets down a bridge.

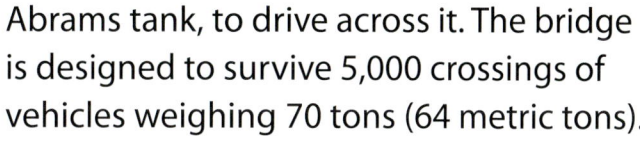
An M104 crosses over the bridge it has placed during a training exercise.

Abrams tank, to drive across it. The bridge is designed to survive 5,000 crossings of vehicles weighing 70 tons (64 metric tons).

BRIDGELAYING

The Wolverine has a crew of two. They are protected inside the armored vehicle while controlling the bridge installation. It takes only about four minutes to lay the bridge down. Picking the bridge back up, which can be done from either end, takes about ten minutes.

M113A3 ARMORED PERSONNEL CARRIER (APC)

M113 armored personnel carriers that are designed for medical use may be marked with a red cross.

The M113A3 armored personnel carrier (APC) is designed to carry soldiers into battle. Eleven soldiers can ride inside. They exit the vehicle through a ramp in the back. Then the M113A3 drives back to safety. Its aluminum armor protects the troops within. The armor can defend against rifle fire, but it can be penetrated by heavy machine guns.

The M113 entered service in 1961. It has seen many upgrades over the years. The M113A1 in 1964 added a more powerful engine. The M113A2 in 1979 improved the engine cooling and suspension. The M113A3 in 1986 introduced an even better engine and stronger armor. More than 32,000 of these APCs have been built for the US Army. The vehicle is also widely sold to foreign militaries.

VERSIONS AND EQUIPMENT

There are multiple versions of the M113A3. They are used for different roles. Some are used for medical evacuation. Others deliver cargo to the front lines. Weapons such as anti-tank missiles and anti-aircraft guns can be added to the APC.

Smoke may obscure the battlefield, so soldiers practice training in these conditions.

M160 ROBOTIC MINE FLAIL

The M160 Robotic Mine Flail is used to clear areas with mines. This small, tracked vehicle has a rotating device in the front. As the device rotates, it slams the ground at high speeds with hammers on chains. This digs several inches into the soil, disabling or destroying buried land mines. Originally designed in Croatia, the M160 was widely used during the Afghanistan War.

The operator controls the M160 using a handheld device. The vehicle transmits video back to the device so that the

Soldiers practice operating the M160 at army bases so they are prepared for high-risk situations.

The M1271 Mine Clearing Vehicle has a similar hammer-and-chain design to the M160.

operator can see where it is going. The person controlling the vehicle can stand at a safe distance, so he or she won't be injured if a mine explodes. Sometimes the operator works from a nearby vehicle, which gives him or her more protection from enemy gunfire.

DESIGN

The M160 is protected by steel armor. It can survive an explosion at close range. The vehicle is designed to work in many environments, including cities, fields, and forests. In addition to the usual flail, it can be equipped with a roller to set off mines or a bulldozer blade to clear an area.

M1070 HEAVY EQUIPMENT TRANSPORTER

The M1A1 Abrams tank was heavier than past US tanks. The Army's trucks were not able to carry it. This problem was solved by the M1070 Heavy Equipment Transporter. The large truck entered service in 1992. The eight-wheeled vehicle is designed to provide a smooth ride over rough terrain when carrying extremely heavy cargo.

DESIGN AND UPGRADES

The M1070 tows Abrams tanks on the M1000 trailer. It also has space in the cabin for the tank's crew. The M1070 truck is able to carry other heavy loads as well. It can transport combat vehicles, howitzers, and construction equipment. The M1070

The M1070 Heavy Equipment Transporter can carry about 70 tons (64 metric tons) of cargo and equipment.

itself can be carried by the Air Force's largest transports, including the C-17 and the C-5.

As with many Army vehicles, the M1070 has seen upgrades over time. The M1070A1 entered service in 2012. It features an improved engine, better brakes, and a new electrical system. The upgrade also adds an armored cab, giving more protection to the crew.

Soldiers guide an Assault Breacher Vehicle onto the trailer of an M1070.

M1126 STRYKER COMBAT VEHICLE

Students drive Stryker Combat Vehicles as part of the Scout Leader Course, which is designed to improve reconnaissance skills.

The M1126 Stryker Combat Vehicle is used by the Army. This eight-wheeled armored vehicle is designed to be lighter and easier to transport than other armored vehicles, such as tanks. It is named for two winners of the Medal of Honor: Stuart S. Stryker, who fought in World War II, and Robert F. Stryker, who fought in the Vietnam War. The Stryker Combat Vehicle entered service in 2002.

ROLE IN THE MILITARY

Strykers fill many different roles. Some carry up to nine troops into battle. Others perform reconnaissance missions. Strykers may be used for medical evacuations. They can also be used to fire anti-tank missiles.

The original Strykers had flat bottoms. Starting with the second generation of the vehicle in 2009, the Stryker gained a double-V-shaped hull to better survive explosions. Third-generation Strykers began entering service in 2020. They included a more powerful engine and larger tires. They were also designed to network with other Army vehicles to quickly share vital information.

The US military has more than 2,100 Stryker Combat Vehicles.

M1150 ASSAULT BREACHER VEHICLE

The M1150 Assault Breacher Vehicle is designed to clear minefields and battlefield obstacles, allowing other vehicles to pass through. The Marine Corps and the Army use it. The M1150 is sometimes called the Breacher or the Shredder.

Two crew members operate an M1150 equipped with a mine plough to clear explosive devices.

The Breacher was first used in combat in 2009, during the Afghanistan War.

DESIGN

The M1150 is based on the M1A1 Abrams tank. Instead of a turret, the M1150 can be equipped with other structures. For example, it can use a plough attachment to push mines out of the way. It can also use a bulldozer attachment to clear obstacles or build defensive positions. The top structure features two launchers for mine-clearing charges. These launchers shoot an explosive rope up to 492 feet (150 m) forward. When the rope explodes, it destroys mines in its path.

Like the Abrams tank, the Breacher has heavy armor that protects the crew inside. The M1150 has a crew of two, with a commander and a driver. The commander can operate a machine gun mounted on the top if needed. The M1150 can also be operated remotely if an area is especially dangerous.

MARINE PROTECTOR–CLASS COASTAL PATROL BOAT

Marine Protector–class boats reach speeds of 25 knots (46 kmh).

The Coast Guard's Marine Protector–class coastal patrol boats have a variety of missions. They find and rescue people in distress. They enforce the law, working against drug trafficking and illegal immigration. They also patrol fishing areas. Out of more than 70 built, four of the patrol boats were made solely to protect Navy submarines while they approach and leave their bases in Georgia and Washington. The Marine Protector class can operate up to 200 miles (320 km) offshore.

These patrol boats are able to function in rough seas. Two powerful diesel engines push them through the water.

The control room features advanced technology, including modern radar and electronic navigation systems.

CREW

The Marine Protector class was designed with more crew comfort compared with past coastal patrol boats. The crew of 11 has access to two bathrooms with showers. The dining area has seating for nine people, and it includes a television and DVD player.

Small boats can be launched from ramps at the back of Marine Protector–class coastal patrol boats.

MEDIUM TACTICAL VEHICLE REPLACEMENT (MTVR)

The Medium Tactical Vehicle Replacement (MTVR) is the main cargo truck for the Marine Corps. The Navy also uses this vehicle. The MTVR entered service in 2001. It replaced older models of trucks, including the M939 and the M809. The MTVR offers better reliability and cargo capacity than past trucks.

> The MTVR is still operational even in wet, snowy conditions.

A Navy equipment operator helps guide an MTVR onto a trailer for transport.

CARRYING CAPACITY

The MTVR is a six-wheeled truck with all-wheel drive for operating on rough terrain. It can carry a maximum of 15 tons (14 metric tons) of cargo. It can also tow another 11 tons (10 metric tons). The MTVR moves the M777 howitzer, troops, fuel, water, and other equipment and supplies. If the truck is operating in dangerous areas, crews can add armor and mount a machine gun on top.

Besides the main model, there are several different versions of the MTVR. Some have extended cargo beds. Others are used as dump trucks or tractors. One type resupplies HIMARS rocket launchers.

MH-65E DOLPHIN

A Coast Guard rescue swimmer does a free fall from a Dolphin.

The MH-65E Dolphin is a short-range rescue helicopter used by the Coast Guard. It holds enough fuel to fly for only about 3.5 hours, but it moves quickly. The Dolphin can fly at speeds of up to 165 knots (310 kmh) in short bursts. It is the standard helicopter used on all Coast Guard cutters with flight decks.

UPGRADES OVER THE YEARS

The helicopter entered service as the H-65 Dolphin in the 1980s. A major upgrade added weapons to the helicopter for its anti-drug mission. An M240 machine gun and a powerful M107 rifle can be used to disable boats carrying illegal drugs. The upgrade also added a night-vision system for carrying out missions in the dark. After these changes, the helicopter became known as the MH-65C.

Further upgrades in the 2010s and 2020s led to the MH-65D and MH-65E. The E model features advanced GPS gear, improved engines, weather radar, and much more.

Coast Guard crew members tie down an MH-65E after it lands on the flight deck of a cutter.

MQ-1C GRAY EAGLE

The MQ-1C Gray Eagle is a remotely piloted aircraft used by the Army. These types of vehicles are commonly known as drones. The Gray Eagle has two missions. First, it uses its radar and sensors to observe the battlefield from above. Second, it uses missiles and bombs to strike targets. Crews control the drone from a ground station.

HISTORY

The Gray Eagle descends from the RQ-1 Predator. This drone entered service with the Air Force in 1996. The *R* in

The Gray Eagle has a wingspan of 56 feet (17 m).

Soldiers inspect an MQ-1C to ensure that the drone is ready for flight.

its name stands for "reconnaissance," and the Q means it is a remotely piloted aircraft. In 2002, the Air Force began using it to fire missiles. Its name changed to MQ-1, with M standing for "multi-role."

In the early 2000s, the Army wanted a drone of its own. It decided to use an upgraded Predator. The Army originally planned to call it the Warrior. The final name became the MQ-1C Gray Eagle. It entered service in 2009.

MQ-9 REAPER

The MQ-9 Reaper is a remotely piloted aircraft flown by the Air Force. The Reaper's main mission is reconnaissance. It uses a set of sensors to keep an eye on the ground. It gathers data about enemy movements, helps search and rescue teams, and watches over friendly forces.

The Reaper can fly at altitudes of more than 50,000 feet (15,240 m).

The Reaper is also able to attack targets. It carries up to eight AGM-114 Hellfire missiles. The aircraft uses a laser targeting system to accurately fire these missiles at enemy vehicles and troops. It can also use this laser to point out targets for other planes to strike. The Reaper has assisted aircraft such as the B-1B Lancer and F-15E Strike Eagle in this way.

GROUND CONTROL

The aircraft has a crew of two people on the ground. One person flies the aircraft, while the other manages the sensors and weapons. The crew is based in the United States. Advanced data links allow them to control the Reaper from across the globe.

The Reaper's advanced sensors help ground crew accurately control the aircraft during takeoff and landing.

MRZR ALPHA LIGHT TACTICAL VEHICLE (LTV)

The MRZR can achieve top speeds of 60 miles per hour (97 kmh).

The MRZR Alpha Light Tactical Vehicle (LTV) is built by the company Polaris, best known for making off-road vehicles for civilian use. The MRZR looks similar to those vehicles. It is designed to quickly move soldiers and light cargo across the battlefield. The vehicle is used by the US Special Operations Command (SOCOM).

DESIGN

The MRZR's small size and light weight mean it can be carried by many types of cargo aircraft, including the MV-22B Osprey tiltrotor and the CH-47 cargo helicopter. The MRZR's large wheels and four-wheel drive help it move around off-road. Soldiers can mount weapons on the vehicle, and it has systems designed to fight enemy drones.

The MRZR can carry two or four people, depending on the version. Some models of the MRZR have open-air cabins. But Polaris has also developed MRZR versions for certain environments. A conversion kit for Arctic missions features an enclosed cabin and replaces the wheels with snowmobile-like tracks.

Marines use a version of the MRZR for amphibious missions.

MV-22B OSPREY

The MV-22B Osprey is a tiltrotor cargo aircraft used by the Marine Corps. It takes off vertically, like a helicopter, using its two large rotors. After takeoff, its rotors tilt forward, transitioning it to fly like an airplane. It returns to vertical flight for landing. The Osprey replaced the CH-46E Sea Knight helicopter. Compared with the Sea Knight, the Osprey provides double the speed, triple the payload, and six times the range. The Osprey can carry up to 32 troops.

An Osprey operated by the Marine Corps prepares for landing in New York City.

The Osprey is crewed by two pilots and one crew chief.

DEVELOPMENT AND DESIGN

The development of the Osprey was long and challenging. The project began in the early 1980s. The aircraft's first flight was in 1989. It finally entered service in 1999. Since then, it has become a capable and reliable aircraft.

Other military branches use different versions of the Osprey. The Air Force began using the CV-22 in 2006. It added additional fuel and an updated radar system. The Navy began flying the CMV-22 in 2020. The vehicle delivers people and cargo to aircraft carriers at sea.

NIMITZ-CLASS AIRCRAFT CARRIER

The Nimitz-class aircraft carriers are the backbone of the Navy's power at sea. The ten carriers in this class are powered by nuclear reactors. They can carry dozens of aircraft at once. This includes fighters like the F/A-18 Super Hornet, command and control aircraft like the E-2 Hawkeye, cargo planes like the C-2 Greyhound, and several helicopters. An angled flight deck allows the carrier to launch and land airplanes at the same time.

Nimitz-class carriers are 1,092 feet (333 m) long.

CREW AND SYSTEMS

The Nimitz class is named for Admiral Chester Nimitz, who led Navy forces in World War II. Nimitz-class ships are enormous, and they are sometimes described as small cities at sea. A crew of about 3,200 people operate the ship. Another 2,500 people fly and support the aircraft. The carrier has dentists and doctors, a post office, a library, barbershops, and stores.

Besides their planes, the carriers have other ways to defend themselves. They can fire anti-aircraft and anti-ship missiles, and they have multiple Phalanx close-in weapon systems (CIWS) to shoot down threats. They can also jam enemy radar and launch decoys to confuse incoming missiles.

Nimitz-class aircraft carriers are durable ships designed to serve for about 50 years.

OHIO-CLASS SUBMARINE

Many Ohio-class submarines are named after US states, such as the USS *Alaska*, which uses ballistic missiles.

Ohio-class submarines were built between 1974 and 1997. Nuclear reactors power these Navy submarines, which were designed to launch missiles. Fourteen of these submarines launch ballistic missiles. These missiles carry nuclear weapons to distant targets. The other four Ohio-class submarines were modified to launch nonnuclear Tomahawk missiles instead.

The ballistic missile submarines are the only US military vehicles that use the Trident II missile. Each submarine holds 20 of these missiles, which can be launched from underwater. Steam in the launch tube propels the missile upward. When the

missile breaks the surface, its rocket engine fires. The Trident II has a range of about 4,600 miles (7,410 km). Each missile is able to carry multiple nuclear warheads that can hit different targets.

CREW

An Ohio-class submarine has a crew of about 150 people. Each submarine has two crews, Blue and Gold. They alternate patrols, meaning that a submarine always has a well-rested crew ready to go when it returns to shore.

The crew of the USS *Tennessee*, an Ohio-class submarine, returns to shore.

P-8A POSEIDON

The P-8A Poseidon is a Navy maritime patrol aircraft. Its main mission is anti-submarine warfare. The Poseidon has advanced radar and other sensors to help it detect and track enemies. It can drop floating devices called sonobuoys into the water. Sonobuoys use sonar, emitting sound and listening for the reflections to locate underwater objects, including submarines. The sonobuoys

The Poseidon can perform both high-altitude and low-altitude missions.

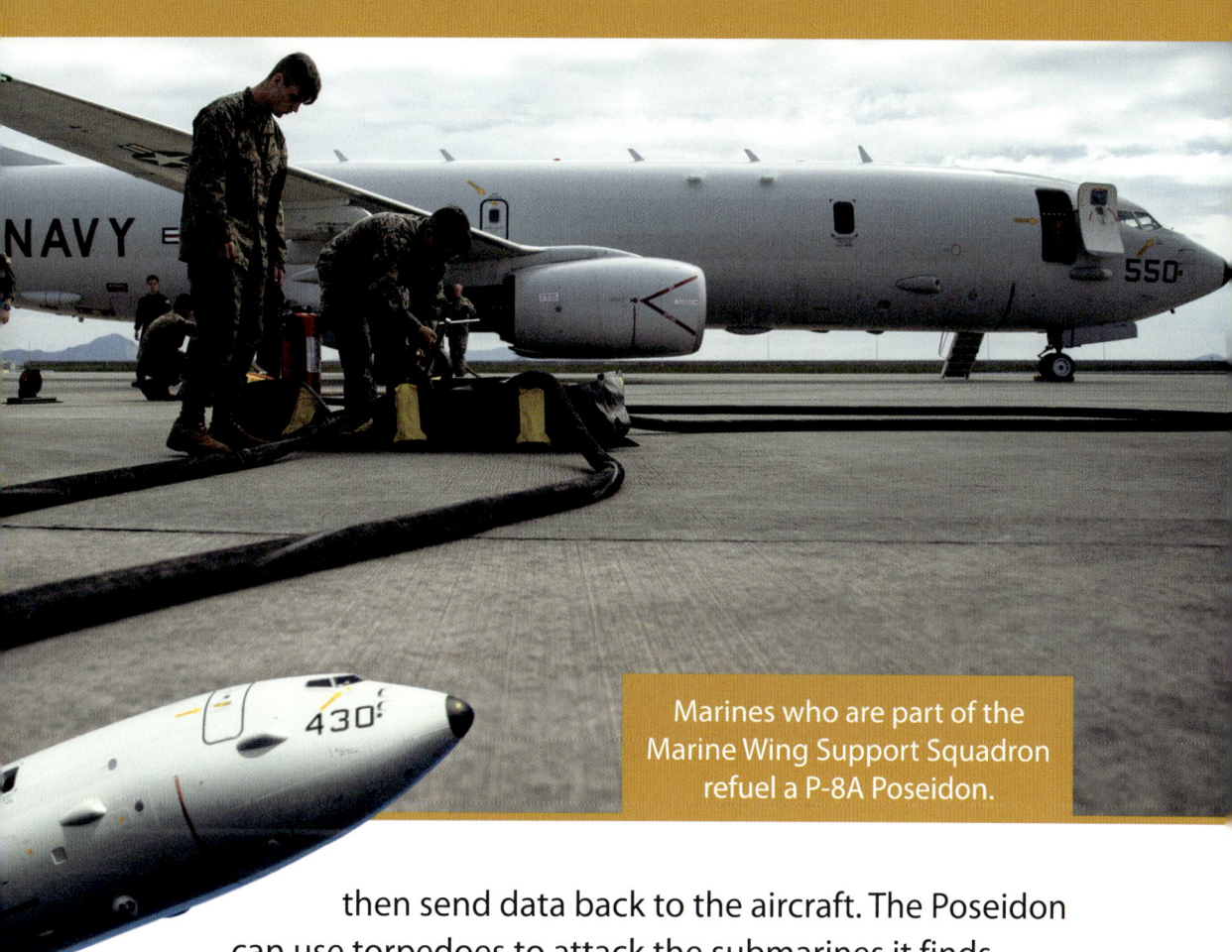

Marines who are part of the Marine Wing Support Squadron refuel a P-8A Poseidon.

then send data back to the aircraft. The Poseidon can use torpedoes to attack the submarines it finds.

DEVELOPMENT AND DESIGN

The P-8A was developed from the Boeing 737 airliner. It entered service in 2013. The aircraft has a crew of nine people. In addition to fighting submarines, it can be used to attack surface ships and carry out reconnaissance missions. The Poseidon can also use its sensors to help with search and rescue missions. It can release a survival kit containing a raft and other supplies to help people stranded at sea.

R-11 AIRCRAFT REFUELER

Labels warn people not to smoke near R-11 aircraft refuelers, as their contents are extremely flammable.

The R-11 truck is the main refueling vehicle used by the Air Force. The first version went into service in 1987, replacing the earlier R-9. Newer generations of the refueler arrived in 1994 and 2004.

The six-wheeled truck has a large aluminum tank that holds up to 6,070 gallons (23,000 L) of jet fuel. It can pump 605 gallons (2,300 L) of fuel per minute. The pumping equipment sits between the tank and the two-door cab, which seats a driver and one passenger.

CREW

The R-11 is designed to quickly get aircraft back into the sky. Crews position the refueler near the runway. A returning aircraft can park near the R-11, and the crew can refuel the jet while its engine is still running. Ground crews train and run drills to make this process as efficient and safe as possible.

Heavy-duty hoses transfer fuel into and out of the R-11 refueler.

RELIANCE-CLASS CUTTER

The Reliance-class cutters of the Coast Guard are Medium Endurance Cutters, similar to the Famous-class vessels.

Small boats can be deployed at sea from Reliance-class cutters.

Reliance-class cutters are 210 feet (64 m) long.

Their main missions are law enforcement and search and rescue. Each cutter has a crew of 77 people. There is no hangar for storing aircraft inside, but the Reliance class can operate one MH-65E Dolphin from the flight deck. The cutter features a Mk-38 cannon and two machine guns.

CHANGES OVER TIME

The first Reliance-class cutter entered service in 1964. It offered better crew comfort compared with past vessels, with air conditioning in most indoor spaces. The pilothouse is located high on the vessel and has windows all around, providing excellent visibility of the surroundings. Starting in 2005, all the cutters in this class went through an upgrade process. Major maintenance was completed, and much of the equipment was modernized. This work was finished in 2014.

RESPONSE BOAT-MEDIUM (RB-M)

Coast Guard vessels less than 65 feet (20 m) long are called boats rather than cutters. The Response Boat-Medium (RB-M) is an aluminum boat that is 45 feet (14 m) long. Two diesel engines power a water jet that propels the boat forward. This is safer compared with a propeller when the boat's crew is rescuing someone from the water, a common mission for the RB-M.

DESIGN AND CREW

The RB-M is designed to handle rough seas and high winds. The seats have built-in shock absorbers to help keep the crew comfortable in

A Response Boat-Medium can reach speeds of 30 knots (56 kmh).

A Response Boat-Medium is able to make sharp turns without capsizing.

choppy waters. The boat's edges have a bumper, allowing the RB-M to move alongside another boat without causing damage if the vessels knock together. If the RB-M does capsize, it is able to safely return itself to an upright position.

The boat has a crew of four. Crew members can mount weapons at the front and rear of the boat for law-enforcement missions. The RB-M also has room for up to five people rescued from the water.

RESPONSE BOAT-SMALL (RB-S) II

The Coast Guard's Response Boat-Small (RB-S) entered service in 2003. The next generation of these vessels, the RB-S II, was introduced in 2012. The Coast Guard purchased a total of 370 of these boats. The last entered service in 2019.

ROLE IN THE COAST GUARD

The RB-S II features better crew comfort than the earlier model. Shock-absorbing seats help keep the four-person crew steady when navigating rough waters. There are

It takes skill and training to maneuver small, fast boats in close proximity to one another.

Members of the Coast Guard practice handling the RB-S II at high speeds.

mounts for weapons in the front and back. The front weapon seat is also shock-absorbing. The cabin is lightly armored to protect against gunfire, and the windows can be opened for ventilation.

Like other Coast Guard vessels, the RB-S II is involved in port security, search and rescue, and law enforcement missions. Crews also use it to board vessels. The boat is 28.7 feet (8.7 m) long. Its twin outboard motors give it a top speed of more than 40 knots (74 kmh). The RB-S II has a range of about 150 nautical miles (280 km).

RQ-4 GLOBAL HAWK

The RQ-4 Global Hawk is a large remotely piloted aircraft used by the Air Force. Like other drones, its main mission is surveillance. It carries advanced radar and many other sensors to study the landscape below. A telescope gives its cameras a close-up view of objects on the ground.

The RQ-4 has a wingspan of 131 feet (40 m) and flies for long distances at high altitudes. The Global Hawk can remain airborne for more than 30 hours without refueling. It soars at an altitude of 65,000 feet (19,810 m), which makes it difficult for missiles to hit. The airplane also has a jamming system to counter missiles.

The Global Hawk is the largest drone used by the US military.

Airmen who remotely control drones such as the RQ-4 Global Hawk require special training.

CREW

Different crews and pieces of equipment are needed to operate a Global Hawk during various stages of flight. The Launch and Recovery Element (LRE) controls the aircraft during takeoff and landing. It includes one pilot and the communications gear. The Mission Control Element (MCE) flies the rest of the mission. It includes a different pilot, a sensor operator, and all of the equipment they need.

RQ-11B RAVEN

The RQ-11B Raven is a small remotely piloted aircraft. It is widely used in the military. Troops in the Army, the Air Force, the Marine Corps, and special operations forces use it. It entered service in 2004 and

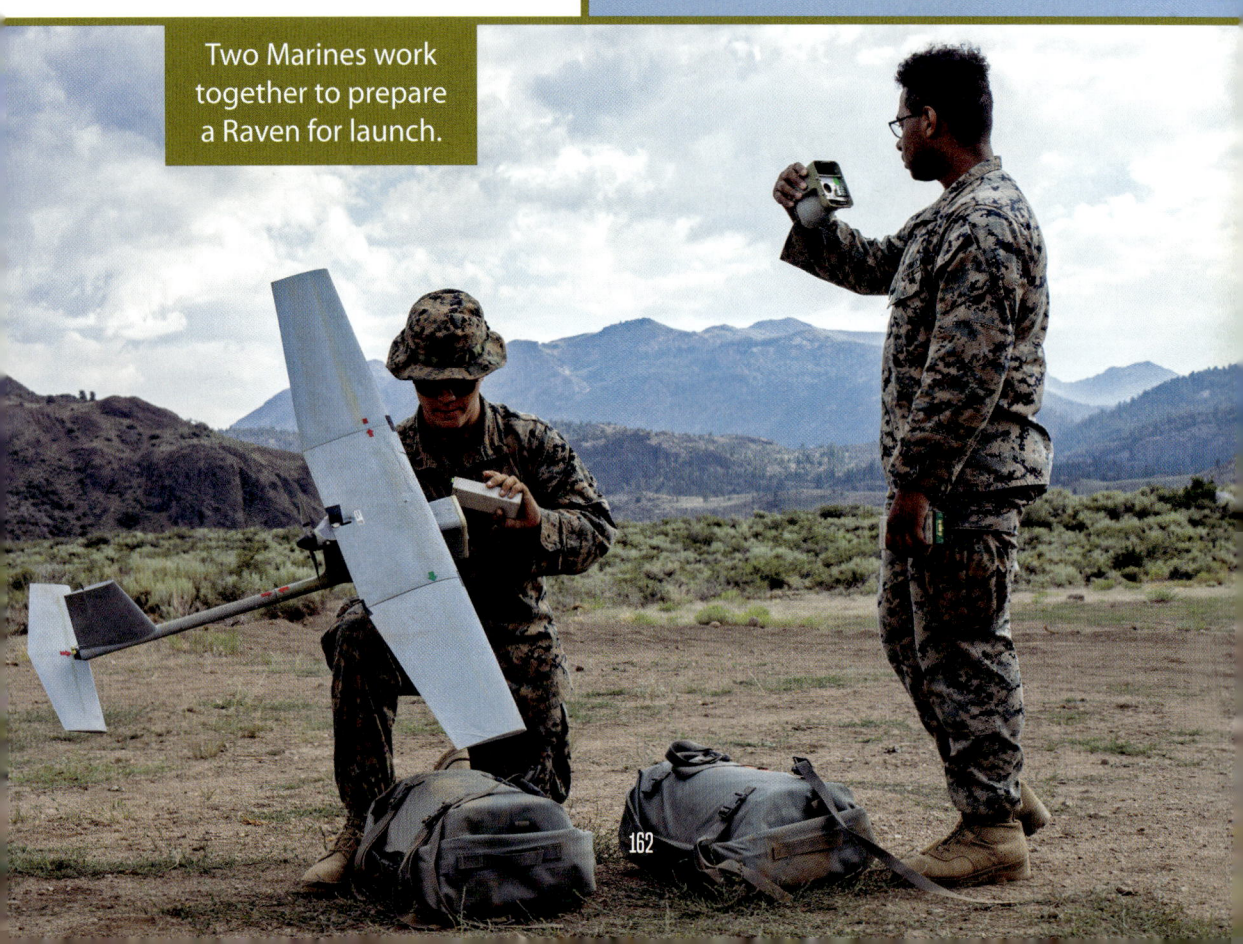

Two Marines work together to prepare a Raven for launch.

An Army soldier releases an RQ-11 Raven.

saw use in the Iraq War (2003–2011) and the Afghanistan War.

The Raven has a wingspan of just 4.5 feet (1.4 m). It weighs less than 5 pounds (2.3 kg). Troops launch it by throwing it into the air. The battery-powered plane can fly for up to 90 minutes. Soldiers can either steer it directly or program it to follow a planned path. The Raven does not need a runway to land. It simply flies down to the ground slowly enough that it can come to rest without damage.

VIDEO SYSTEMS

Three cameras on the drone collect video footage from high above the ground. This video is transmitted back to the ground control station. It takes two people to operate the Raven.

RQ-20B PUMA

The RQ-20B Puma is a small remotely piloted aircraft. Its cameras provide valuable intelligence to troops on the ground or on Navy ships. It is larger than the RQ-11B Raven, with a wingspan of 9.2 feet (2.8 m) and a weight of 14 pounds (6 kg). Its battery allows the Puma to fly for more than three hours, covering a maximum range of about 12.4 miles (20 km). It flies at an altitude of about 500 feet (150 m).

A soldier inspects a Puma before launch.

In the early 2020s, the Army announced plans to expand its use of the Puma.

OPERATION

Operators can launch the Puma either by hand or by using a launch rail. It can be steered from the ground or follow a programmed route. The operators use the same type of ground control station as with the Raven.

The company AeroVironment, which also makes the Raven, originally designed the Puma for special operations forces. The Navy, the Marine Corps, and the Army later began using the Puma too. The aircraft can fly down for a gentle landing on the ground or in the water. It can also land in a net on a ship.

SENTINEL-CLASS CUTTER

The Coast Guard's Sentinel-class cutters carry out the branch's missions. These include security in ports and waterways, law enforcement, and search and rescue. The ship's sensors and communications gear are designed to work seamlessly with other cutters and with Navy ships.

Sentinel-class cutters are designed to quickly respond to emergencies, reaching speeds of 28 knots (52 kmh).

In 2022, the USCGC *William Chadwick* became the 50th cutter to join the Sentinel-class fleet.

The Sentinel class entered service in 2012. Each cutter is named after a notable enlisted member of the Coast Guard. For example, the first cutter was named the USCGC *Bernard C. Webber*. It honors a Coast Guardsman who led a daring rescue mission of shipwrecked sailors off the coast of Massachusetts in 1952.

CREW

Sentinel-class cutters have a crew of 25 people. The crew can launch a small boat off the rear of the ship for rescue missions. They can also use several weapons for self-defense. A remotely operated Mk-38 Bushmaster autocannon near the front provides heavy firepower. There are also four mounted machine guns that crew members can fire.

SPACE-BASED INFRARED SYSTEM (SBIRS) SATELLITES

The Space-Based Infrared System (SBIRS) satellites are operated by the Space Force. While in orbit, these satellites watch for missile launches on Earth. They contain sensors that detect the heat from a launch. Then they send this data back to the Space Force. SBIRS can tell what kind of missile is launching, the direction in which it is going, and where it is likely to land.

SATELLITE DESIGN

SBIRS includes six satellites in geosynchronous orbit. This means they orbit at the same speed as Earth's rotation, keeping them over the same region.

An Atlas V rocket is transported to a launch site. Rockets are necessary to launch satellites into space.

A rocket carrying an SBIRS satellite successfully launched from Cape Canaveral in Florida in 2017.

Together the SBIRS satellites have global coverage. The satellites take images every ten seconds to search for the infrared signature of a missile launch anywhere on Earth.

The satellites use solar panels to generate power. With their panels outstretched, the satellites are about 49 feet (15 m) wide. Each one contains about 1,100 pounds (500 kg) of sensor equipment. The satellites are launched with 430 pounds (195 kg) of fuel to help maintain their orbits for many years.

T-6A TEXAN II

The T-6A Texan II is a training aircraft used by the Air Force and Navy. New pilots fly the propeller-driven plane to learn and practice basic flying skills. From there, they can go on to fly fighters, bombers, cargo planes, or helicopters for the Air Force or Navy.

MOBILE MACHINE

The T-6A is based on the Pilatus PC-9, a Swiss training aircraft. The T-6A entered service in the United States in 2000. The plane's single engine can bring it to an altitude of 31,000 feet

A student and an instructor prepare for takeoff in a T-6 Texan II.

T-6 Texan IIs are small aircraft with a wingspan of just 33.4 feet (10.2 m).

(9,450 m) and speeds of 320 miles per hour (515 kmh). It can perform rolls, loops, and other aerobatic maneuvers, preparing pilots for faster and more agile aircraft.

The plane has two seats, one for the student and one for the instructor. The aircraft can be flown from either seat. The plane includes ejection seats for emergencies.

T-38C TALON

The T-38C Talon is a supersonic jet trainer flown by the Air Force and Navy. The twin-engine jet helps prepare pilots to fly high-performance fighters. It has two seats, with the student in front and the instructor in back. The seats have ejection devices in case of emergency.

DEVELOPMENTS OVER THE YEARS

The original T-38 dates back to 1961. It was the first training aircraft to fly faster than the speed of sound. Since then, more than 70,000 Air Force pilots have trained in these jets. Pilots from allied countries come to the United States to train in the T-38. NASA astronauts also use the T-38 for flight training.

The T-38C model first flew in 2001. This updated trainer features more reliable engines along with modern displays and controls. Another version is the AT-38B. It includes a practice bomb dispenser so pilots can train for ground attacks.

The T-38 Talon can fly at high altitudes above 55,000 feet (16,760 m).

In addition to training students, the T-38 Talon can be used to see how well two pilots work together.

TICONDEROGA-CLASS CRUISER

The Navy's Ticonderoga-class cruisers carry out many kinds of missions. They can fight aircraft, submarines, and surface ships. Sometimes they join other vessels in a carrier battle group. Other times they work independently. Each ship has a crew of about 360 people.

WEAPONS SYSTEMS

The cruiser's main weapons are guided missiles. They use the Mk-41 Vertical Launching System to shoot anti-air missiles,

Twenty-seven Ticonderoga-class cruisers were built between 1983 and 1994.

Ticonderoga-class cruisers can travel at speeds of more than 30 knots (56 kmh).

anti-submarine missiles, and Tomahawk cruise missiles. The ships also have torpedoes and Mk-45 naval guns. They carry two SH-60 Seahawk helicopters for anti-submarine warfare. The cruisers also have two Phalanx CIWS for self-defense.

The Ticonderoga class is part of the Navy's Aegis combat system. This system uses advanced radar equipment and computers to track, target, and destroy enemy ships and aircraft. A few of the Ticonderoga-class ships also have the Aegis missile defense system. This equipment shoots down long-range ballistic missiles before they can reach the United States. It fires the SM-3 missile to intercept these weapons.

UH-1Y VENOM

The UH-1Y Venom is a utility helicopter in service with the Marine Corps. It carries out many types of missions. The Venom can deliver cargo to ships at sea. It can transport Marines into battle. The crew can fly search and rescue or reconnaissance missions.

The Venom has a crew of four people: a pilot, a copilot, a crew chief, and a gunner. Helmet-mounted displays keep important information in view of the pilot and copilot at all times. The crew chief helps maintain the helicopter. The gunner operates powerful machine guns mounted in the doors. The helicopter can also carry eight Marines along with their gear.

Gunners practice firing at targets from within UH-1Y helicopters.

A Navy sailor is hoisted into a UH-1Y Venom helicopter.

UPGRADES AND PARTS

The Venom is an upgraded version of the earlier UH-1N Huey helicopter. It features improved engines and rotors, carries a greater payload, and adds modern cockpit displays. The Venom has many parts in common with the AH-1Z Super Cobra. This makes it easier and cheaper for the Marine Corps to maintain and repair both helicopters.

UH-60M BLACK HAWK

The UH-60M Black Hawk is a utility helicopter in service with the Army. It delivers troops to the front lines, evacuates wounded soldiers, and supports special operations missions. There is room inside for 11 people. The helicopter can carry up to 9,000 pounds (4,080 kg) of cargo either inside or underneath using a sling system.

DESIGN

The Black Hawk is a rugged helicopter. Armored floors protect against gunfire from the ground. Backup systems keep the helicopter flying if important parts get damaged. If the Black Hawk does crash, its structure is designed to crumple on impact. This absorbs some of the impact of the crash, protecting the people within.

Worldwide, there are more than 4,000 Black Hawk aircraft in operation. The US Army operates the most of any military branch in the world, with about 2,135 of the helicopters.

A Black Hawk helicopter is crewed by a pilot, copilot, and crew chief.

The original UH-60A Black Hawk entered service with the Army in 1979. It has been upgraded several times over the years. The UH-60M arrived in 2007. It has improved electronic equipment and more powerful engines compared with past models.

VIRGINIA-CLASS ATTACK SUBMARINE

The crew of the USS *Indiana*, a Virginia-class attack submarine, saluted as their vessel was commissioned into service in 2018.

The Navy's Virginia-class submarines are nuclear-powered boats designed to attack enemy submarines and surface ships. They also have launch tubes for firing Tomahawk cruise missiles. The submarines may assist special operations forces. They can deliver a small group of soldiers to an enemy shoreline without being detected.

The Navy originally planned to replace the Los Angeles–class submarines with Seawolf-class submarines in the 1990s. But the Seawolf class, though very capable, was extremely expensive. The Navy built just three of them. It then developed the Virginia class as a smaller and cheaper alternative. The first Virginia-class submarine arrived in 2004.

CREW AND DESIGN

Most of the Virginia-class submarines are named for US states. Each one has a crew of 132 people. Rather than a traditional periscope, the submarines use a set of cameras and sensors to give the crew a view above the surface. The submarines are designed to be easy to upgrade as technology advances, keeping them in service for decades to come.

Virginia-class attack submarines are designed to be stealthy and fast, capable of traveling at speeds of more than 25 knots (46 kmh).

WIDEBAND GLOBAL SATCOM (WGS) SATELLITES

The Wideband Global SATCOM (WGS) satellites provide communications for the US military. The Space Force develops, launches, and operates these satellites. Space Force personnel manage the overall system from five operations centers on Earth. People on land, at sea, or in the air can use terminals with antennas to connect to the network. The satellites keep them connected with other US military forces around the globe.

WGS satellites are designed to last at least 14 years.

LAUNCHING SATELLITES

The first WGS satellite launched in 2007. In 2023, there were ten satellites in orbit and more planned for the future. The satellites launch on Atlas V or Delta IV rockets. WGS satellites are in geosynchronous orbits. Each satellite covers a particular region of Earth. For example, the WGS-1 satellite is over the Pacific Ocean, and the WGS-2 satellite operates over the Indian Ocean. The satellites use large solar panels to generate the electricity needed to run their communications gear.

WGS-10 launched on a Delta IV rocket in 2019.

X-37B ORBITAL TEST VEHICLE

The X-37B Orbital Test Vehicle is an experimental spacecraft flown by the Space Force. It is remotely controlled. The X-37B is a reusable

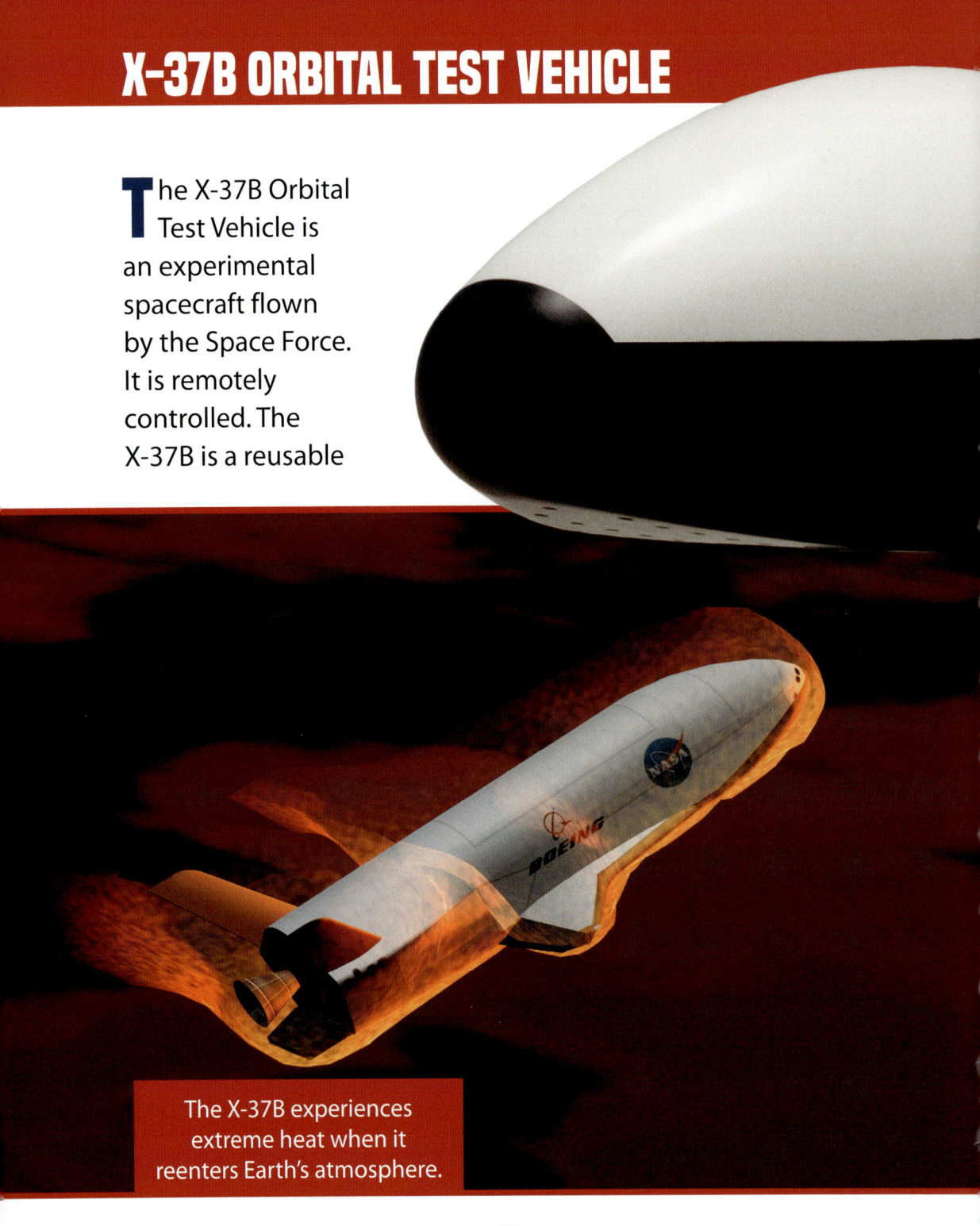

The X-37B experiences extreme heat when it reenters Earth's atmosphere.

The X-37B is designed to function in low Earth orbit, which is just 150 to 500 miles (241–805 km) above Earth.

spacecraft with wings. It launches vertically on a rocket and orbits Earth. Then it returns for a landing on a runway.

ADVANCED TECHNOLOGY

The X-37B is used to test materials and equipment used in spacecraft. It also carries scientific experiments to run in orbit. Many things about the X-37B remain military secrets.

NASA had plans for a spacecraft called the X-37, but this vehicle was never built. The military created the X-37B based on this initial design. The X-37B completed its first mission in 2010. It can remain in space for hundreds of days at a time. In 2022, the spacecraft returned to Earth after 908 days in orbit. It was the vehicle's longest mission yet.

ZUMWALT-CLASS DESTROYER

The Navy's Zumwalt-class destroyers are made to fight enemy aircraft, submarines, and surface ships. The Zumwalt class is named for former Chief of Naval Operations Elmo R. Zumwalt. The first ship entered service in 2016, and each one has a crew of 197 people.

The destroyer includes stealth technology, giving it a unique look. The hull is designed to leave less wake than other warships. A special system reduces the heat given off by the engines. The ship's angled shape helps it hide from radar.

The ship is also designed with technological advancements in mind. Its power system makes more electricity than the ship currently needs. That way it will be able to use next-generation sensors and future high-energy weapons, such as lasers.

Zumwalt-class destroyers are 610 feet (186 m) long.

The USS *Zumwalt* fires a missile as part of a training exercise.

MISSILES

The class features 80 launch cells for firing Tomahawk missiles, anti-air missiles, and anti-submarine missiles. The Navy also designed the ship to fire the Conventional Prompt Strike (CPS) missile. This weapon can travel at extremely high speeds, up to 1 mile (1.6 km) per second, to strike key targets very quickly.

GLOSSARY

ambush
A planned sneak attack.

cargo
Supplies carried by a vehicle.

class
A specific design of ship.

collateral damage
Unintended harm to people or objects near a target.

countermeasures
Actions taken against an enemy threat.

cruise missile
A missile that flies a long distance at a steady speed and precisely hits a target.

cutter
The term used by the Coast Guard for its vessels larger than 65 feet (20 m).

drag
The force of air resistance that slows down moving objects.

hangar
An enclosed space where planes and helicopters are stored.

knot
A unit of speed defined as one nautical mile per hour, equal to 1.15 miles per hour (1.85 kmh).

maneuver
The controlled movement of a vehicle by its operator. Vehicles that are highly maneuverable are agile and have responsive steering.

minigun
A powerful, rapid-firing machine gun with multiple rotating barrels.

nuclear reactor
A device that generates power by splitting apart atoms.

radar
A system that sends out radio waves and then listens for the reflections, allowing it to locate distant objects.

reconnaissance
Exploring an area and sending back information about what is there.

thrust
The force that an aircraft's engines generate to push the vehicle forward or upward.

turret
The rotating structure that holds the main gun on top of a tank or other armored vehicle.

winch
A device used to pull or lift something using a wire, rope, or cable.

TO LEARN MORE

FURTHER READINGS

Marcovitz, Hal. *Military Drones*. ReferencePoint, 2021.

Mooney, Carla. *US Air Force*. Abdo, 2021.

Ringstad, Arnold. *The Military Weapons Encyclopedia*. Abdo, 2024.

ONLINE RESOURCES

To learn more about US military vehicles, please visit **abdobooklinks.com** or scan this QR code. These links are routinely monitored and updated to provide the most current information available.

INDEX

Aegis combat system, 175
aerial refueling, 29, 94–99
Afghanistan War, 29, 44, 86, 124, 163
aircraft carriers, 5, 22, 46, 63, 64, 74–75, 145, 146, 174
amphibious vehicles, 5, 8–9, 18–19, 22, 103
artillery, 35, 69, 116, 135
attack aircraft, 6–7, 10–17, 22, 57, 64–65

ballistic missiles, 148–149, 175
Blue Angels, 67
boats, 71, 106, 132–133, 156–159, 167
bombers, 4, 26–27, 28–31, 170
bombs, 6, 22, 27, 28, 30, 56, 65, 138, 172
buoy tenders, 92–93, 100–101

cannons, 10–11, 16, 22, 55, 58, 65, 155, 167
civil service mariners, 107
cruisers, 5, 174–175
cutters, 5, 70–71, 92–93, 100–101, 104–105, 136, 154–155, 166–167

destroyers, 5, 20–21, 186–187
drones, 4, 17, 75, 89, 138–141, 160–165

Earth orbit, 76, 168–169, 182, 184–185
ejection seats, 171, 172
electronic warfare, 52–53
engineering vehicles, 4, 116–117, 120–121

fighter jets, 4, 22, 52–55, 58–65, 146, 170, 172

fuel, 29, 33, 63, 66, 78, 83, 91, 94–99, 135, 136, 145, 152–153, 169

global positioning system (GPS), 76–77, 100, 137

helicopters, 5, 12–17, 18, 21, 37, 41, 42–43, 51, 71, 73, 80, 85, 89, 104, 106, 136–137, 143, 144, 146, 170, 175, 176–179
helmet-mounted displays, 12, 62, 176

icebreakers, 80–81
infantry fighting vehicles, 4, 114–115, 117
Iraq War, 44, 163

Korean War, 106

laser targeting, 17, 57, 140
law enforcement, 5, 70–71, 78, 93, 132, 155, 157, 159, 166
littoral combat ships (LCS), 72–73, 88–89

machine guns, 9, 16, 19, 25, 43, 45, 70, 93, 103, 104, 112, 114, 119, 122, 131, 135, 137, 155, 167, 176
mine-resistant ambush protected (MRAP) vehicles, 32–33, 44–45, 90, 108
mines, 4, 24–25, 27, 32, 44, 72, 86, 89, 106, 109, 111, 124–125, 130–131

NASA, 172, 185
nuclear reactors, 74, 111, 146, 148, 180
nuclear weapons, 27, 30, 148–149

Persian Gulf War, 7
Phalanx close-in weapon system (CIWS), 19, 146, 175

radar, 17, 26, 28, 46–51, 52–53, 55, 57, 60, 62, 67, 79, 86, 133, 137, 138, 145, 146, 150, 160, 175, 186

satellites, 5, 76–77, 168–169, 182–183
solar panels, 77, 169, 183
sonar, 25, 150
special forces, 14, 106, 111, 142, 162, 165, 178, 180
stealth, 28, 60–61, 62, 186
submarines, 5, 20–21, 72, 89, 110–111, 132, 148–149, 150–151, 174–175, 180–181, 186–187
surveillance, 50, 78, 105, 160

tanker aircraft, 94–99
tanks, 4, 6–7, 13, 16, 37, 39, 112–113, 116, 119, 120–121, 126, 131
tiltrotor aircraft, 143, 144–145
torpedoes, 20–21, 111, 151, 175
trainer aircraft, 170–172
transport aircraft, 4, 10, 36–43
trucks, 4, 37, 68–69, 82–85, 108–109, 126–127, 134–135, 152–153

Vietnam War, 10, 112, 128

World War II, 20, 74, 93, 106, 128, 146

PHOTO CREDITS

Cover Photos: US Navy/DVIDS, front (E-2C Hawkeye, ship, submarine); US Navy, front (Humvee); US Air Force/DVIDS, front (B-1B Lancer), back (left); Chip Somodevilla/Getty Images News/Getty Images, front (MRAP vehicle); US Air Force, front (satellite); US Army, front (M1126 vehicle); Antonio More/DVIDS, front (T-6B Texan II planes); US Army/DVIDS, front (helicopter); NASA, front (X-37); US Air National Guard/DVIDS, back (right)
Interior Photos: US Navy/DVIDS, 1, 5 (bottom), 9, 18–19, 19, 20–21, 21, 22–23 (top), 24, 24–25, 46, 46–47, 53, 64, 65, 66–67, 72–73, 73, 74–75, 75, 88, 104–105, 107, 110–111, 134–135, 146–147, 147, 148–149, 149, 150–151 (bottom), 164–165, 174, 175, 181, 186–187, 187; US Air Force/DVIDS, 3, 6–7, 10, 11, 26–27, 27, 28, 30–31, 36–37, 40–41 (top), 40–41 (bottom), 48–49, 49, 56–57, 57, 94–95 (top), 96, 96–97, 98, 140–141, 141, 152, 153, 160–161 (top), 160–161 (bottom), 169, 170–171 (bottom), 172–173 (top), 172–173 (bottom); US Navy, 4, 52, 106–107, 110, 176–177, 180; US Air Force, 5 (top), 29, 31, 36, 55, 62, 77, 94–95 (bottom), 99; US Air National Guard, 7; US Marine Corps/DVIDS, 8–9, 12, 13, 15, 22–23 (bottom), 34, 35, 44, 63, 102, 102–103, 108–109, 109, 116–117 (top), 116–117 (bottom), 118, 118–119, 134, 137, 143, 150–151 (top), 162, 176; US Army/DVIDS, 14–15, 16, 42, 43, 86–87, 115, 120–121 (bottom), 122, 125, 127, 128, 138, 162–163, 178–179 (top); US Army, 17, 69, 84, 85, 113, 114–115, 129; Chip Somodevilla/Getty Images News/Getty Images, 32; US Army National Guard/DVIDS, 32–33, 131, 164; Airman 1st Class Sara Hoerichs/DVIDS, 38–39 (top); NASA/DVIDS, 38–39 (bottom); US Department of Defense/DVIDS, 45; Master Sgt. Jeremy Lock/DVIDS, 50, 60–61, 61; US Air National Guard/DVIDS, 50–51, 142; Oregon Air National Guard/DVIDS, 54; Master Sgt. Jason Rolfe/DVIDS, 58–59; Staff Sgt. Sheila deVera/DVIDS, 59; Eliyahu Yosef Parypa/Shutterstock Images, 66; Capt. Joseph Warren/DVIDS, 68; US Coast Guard/DVIDS, 70, 70–71, 78–79, 79, 80–81, 81, 92–93, 93, 100–101, 101, 105, 132, 133, 154, 154–155, 156, 157, 158, 158–159, 166, 166–167; Scott Prater/DVIDS, 76; Ohio Army National Guard/DVIDS, 82–83; Bryan Araujo/DVIDS, 83; US National Guard/DVIDS, 87; Stocktrek Images/Getty Images, 88–89; US Marine Corps, 90, 91, 144, 144–145; Idaho Army National Guard/DVIDS, 112–113; Staff Sgt. Jason Ragucci/DVIDS, 120–121 (top); US Army Reserve, 123; Maj. Dan Marchik/DVIDS, 124; Capt. Fernando Ochoa/DVIDS, 126–127; Cpl. Alisha Grezlik/DVIDS, 130–131; US Coast Guard, 136; Sgt. William Begley/DVIDS, 138–139; US Space Force/DVIDS, 168; Antonio More/DVIDS, 170–171 (top); US Department of Defense, 178–179 (bottom); Michael Pierson/DVIDS, 182; Van Ha/DVIDS, 183; NASA, 184, 184–185

ABDOBOOKS.COM

Published by Abdo Reference, a division of ABDO, PO Box 398166, Minneapolis, Minnesota 55439. Copyright © 2024 by Abdo Consulting Group, Inc. International copyrights reserved in all countries. No part of this book may be reproduced in any form without written permission from the publisher. Encyclopedias™ is a trademark and logo of Abdo Reference.

102023
012024

Editor: Walt K. Moon
Series Designer: Colleen McLaren
Production Designers: Cynthia Della-Rovere and Michael J. Williams

LIBRARY OF CONGRESS CONTROL NUMBER: 2023939634

PUBLISHER'S CATALOGING-IN-PUBLICATION DATA

Names: Ringstad, Arnold, author.
Title: The military vehicles encyclopedia / by Arnold Ringstad
Description: Minneapolis, Minnesota: Abdo Reference, 2024 | Series: US military encyclopedias | Includes online resources and index.
Identifiers: ISBN 9781098293048 (lib. bdg.) | ISBN 9798384910985 (ebook)
Subjects: LCSH: Vehicles, Military--Juvenile literature. | Transportation, Military--Juvenile literature. | Military history--Juvenile literature. | United States--Armed Forces--History--Juvenile literature. | Encyclopedias and dictionaries--Juvenile literature.
Classification: DDC 355.83--dc23